The Next Generation Innovation in IoT and Cloud Computing with Applications

The Next Generation Innovation in IoT and Cloud Computing with Applications is a thought-provoking edited book that explores the cutting-edge advancements and transformative potential of the Internet of Things (IoT) and cloud computing. This comprehensive volume brings together leading experts and researchers to delve into the latest developments, emerging trends, and practical applications that define the next era of technological innovation.

Readers will gain valuable insights into how IoT and cloud computing synergize to create a dynamic ecosystem, fostering unprecedented connectivity and efficiency across various industries. The book covers a wide spectrum of topics, including state-of-the-art technologies, security and privacy considerations, and real-world applications in fields such as healthcare, smart cities, agriculture, and more.

With a focus on the future landscape of technology, this edited collection serves as a pivotal resource for academics, professionals, and enthusiasts eager to stay at the forefront of the rapidly evolving IoT and cloud computing domains. By offering a blend of theoretical perspectives and hands-on experiences, *The Next Generation Innovation in IoT and Cloud Computing with Applications* serves as a guide to the forefront of technological progress, providing a roadmap for the exciting possibilities that lie ahead in this era of connectivity and digital transformation.

The Next Generation Innovation in IoT and Cloud Computing with Applications

Edited by

Abid Hussain, Ahmed J. Obaid, Garima Tyagi, and Amit Sharma

CRC Press is an imprint of the
Taylor & Francis Group, an **informa** business

Designed cover image: © Shutterstock

First edition published 2025
by CRC Press
2385 NW Executive Center Drive, Suite 320, Boca Raton FL 33431

and by CRC Press
4 Park Square, Milton Park, Abingdon, Oxon, OX14 4RN

CRC Press is an imprint of Taylor & Francis Group, LLC

Library of Congress Cataloging-in-Publication Data
Names: Obaid, Ahmed J. (Ahmed Jabbar), editor. | Tyagi, Garima, editor. | Hussain, Abid, editor. | Sharma, Amit, (Computer scientist), editor.
Title: The next generation innovation in IoT and cloud computing with applications / edited by Abid Hussain, Ahmed J. Obaid, Garima Tyagi, and Amit Sharma.
Description: First edition. | Boca Raton : CRC Press, 2025. | Includes bibliographical references.
Identifiers: LCCN 2024010389 (print) | LCCN 2024010390 (ebook) | ISBN 9781032524450 (hbk) | ISBN 9781032524481 (pbk) | ISBN 9781003406723 (ebk)
Subjects: LCSH: Internet of things—Forecasting. | Cloud computing—Technological innovations.
Classification: LCC TK5105.8857 .N496 2025 (print) | LCC TK5105.8857 (ebook) | DDC 004.67/8—dc23/eng/20240718
LC record available at https://lccn.loc.gov/2024010389
LC ebook record available at https://lccn.loc.gov/2024010390

ISBN: 978-1-032-52445-0 (hbk)
ISBN: 978-1-032-52448-1 (pbk)
ISBN: 978-1-003-40672-3 (ebk)

DOI: 10.1201/9781003406723

Typeset in Sabon
by Apex CoVantage, LLC

Contents

Preface

In the dynamic landscape of technology, where innovation is the driving force behind progress, the convergence of Internet of Things (IoT) and cloud computing stands as a testament to the transformative power of synergy. As we navigate the intricacies of this technological juncture, it becomes increasingly evident that we are at the cusp of a new era—an era where the amalgamation of IoT and cloud computing is poised to redefine the boundaries of what is possible.

The Next Generation Innovation in IoT and Cloud Computing with Applications is a compendium of cutting-edge insights and breakthroughs in the realm of connected devices and cloud-based solutions. This edited volume brings together the collective expertise of leading researchers, academicians, and industry practitioners who have tirelessly explored the vast potential of this amalgamation.

The book unfolds a narrative that extends beyond the theoretical framework, delving deep into the practical applications that are shaping our present and influencing the future. It serves as a comprehensive guide for researchers, practitioners, and students alike, providing a roadmap through the intricate pathways of innovation in IoT and cloud computing.

The chapters contained herein span a spectrum of topics, ranging from the fundamental principles that underpin IoT and cloud computing to advanced applications that are reshaping industries. From smart cities and healthcare to industrial automation and beyond, each contribution encapsulates a unique facet of the transformative landscape that is emerging.

We owe our gratitude to the contributors who have shared their expertise and insights, enriching this compilation with diverse perspectives and multifaceted approaches. Their dedication to advancing the frontiers of knowledge is evident in the depth and breadth of the content presented.

Our aim as editors is to present a resource that not only captures the state of the art in IoT and cloud computing but also inspires further exploration and innovation. The journey through these pages will undoubtedly ignite the imagination of those who seek to harness the full potential of the next generation of technology.

We hope that this book serves as a catalyst for ongoing dialogue, collaboration, and exploration in the realm of IoT and cloud computing. May it spark new ideas, fuel curiosity, and propel us toward a future where innovation knows no bounds.

About the Editors

Abid Hussain is Associate Professor in the School of Computer Applications and Dean, Research and Higher Studies at Career Point University, Kota (Raj.) He received his PhD in computer application. He is a chair- person of IPR Cell at Career Point University, Kota. He has over 16 years' teaching experience of higher education, including Under Graduate and Post Graduate courses. His areas of interest are cloud computing, network security, open source technologies, web mining, web engineering, and cyber security. He is also a research supervisor in computer science and technology at Career Point University. He published more than 30+ research papers in the reputed UGC Care and Scopus Indexed international journals of computer science and technology. He is also working as a reviewer and technical program committee member for various national and international conferences as well as research journals. He has worked as session chair and keynote speaker in various international conferences. He has published 7+ patents on the latest technologies in computer science. He has published 3 authored and 6 edited books in computer science and technology. He also works as an external examiner in various universities for PhD evaluation. He is an active member of World Academy of Science, Engineering and Technology(WASET), International Association of Engineers (IAENG), Computer Science Teachers Association(CSTA), International Computer Science and Engineering Society(ICSES), Institute of Research Engineers & Doctors(theIRED) and International Academy for Science & Technology Education and Research (IASTER).

Ahmed J. Obaid is an Asst. Professor at the Department of Computer Science, Faculty of Computer Science and Mathematics, University of Kufa, Iraq. Dr. Ahmed holds a Bachelor's in Computer Science, a degree in – Information Systems from the College of Computers, University of Anbar, Iraq (2001–2005), and a Master's Degree (M. TECH) in Computer Science Engineering (CSE) from School of Information Technology, Jawaharlal Nehru Technological University, Hyderabad, India (2010–2013), and a Doctor of Philosophy (Ph.D.) in Web Mining from College of Information Technology, University of Babylon, Iraq

(2013–2017). He is a Certified Web Mining Consultant with over 14 years of experience working as a Faculty Member at the University of Kufa, Iraq. He has taught courses in Web Designing, Web Scripting, JavaScript, VB.Net, MATLAB Toolbox, and other courses on PHP, CMC, and DHTML from more than 10 international organizations and institutes in the USA, and India. Dr. Ahmed is a member of the Statistical and Information Consultation Center (SICC), University of Kufa, Iraq. His main line of research is Web mining Techniques and Applications, Image processing in Web Platforms, Image processing, Genetic Algorithm, information theory, and Medical Health Applications. Ahmed J. is an Associated Editor in the Brazilian Journal of Operations & Production Management (BJO&PM) and an Editorial Board member in the *International Journal of Advance Study and Research Work* (IJASRW), *Journal of Research in Engineering and Applied Sciences* (JREAS), *GRD Journal for Engineering* (GRDJE), *International Research Journal of Multidisciplinary Science & Technology* (IRJMST), *The International Journal of Technology Information and Computer* (IJTIC), *Career Point International Journal Research* (CPIJR). Ahmed J. was Editor in Many International Conferences such as: ISCPS_2020, MAICT_2020, IHICPS_2020, IICESAT_2021, IICPS_2020, ICPAS_2021, etc. (Scopus Indexed Conferences). He has edited some books, such as *Advance Material Science and Engineering* (ISBN: 9783035736779, Scientific. net publisher), *Computational Intelligence Techniques for Combating COVID-19* (ISBN: 978-3-030-68936-0 EAI/Springer), *A Fusion of Artificial Intelligence and Internet of Things for Emerging Cyber Systems* (ISBN: 978-3-030-76653-5 Springer). Ahmed J. has supervised several final projects for Bachelor's and Master's in his main line of work and authored and co-authored several scientific publications in journals, Books, and conferences with more than 75+ Journal Research Articles, 5+ book Chapters, 15+ Conference papers, 10+ Conference proceedings, 8+ Books Editing, 2+ Patent. Ahmed J. is also a Reviewer in many Scopus, SCI, and ESCI Journals e.g., CMC, IETE, IJAACS, IJIPM, IJKBD, IJBSR, IET, IJUFKS, and many others. Dr. Ahmed attended and participate as: Keynote Speakers (60+ Conferences), Webinars (10+), and Session Chairs (10+), in many international events in the following countries: India, Turkey, Nepal, Philippines, Vietnam, Thailand, Indonesia, and other countries.

Garima Tyagi is an Associate Professor in the School of Computer Application at Career Point University. She is having 25+ years' experience in Higher Education for UG and PG courses. She received post-graduation degrees in Chemistry from Rohilkhand University and Computer Applications from JNRV University respectively. Completed Executive MBA in HR. She received her PhD. Degree in Computer Applications and Technology. Her research area are VOIP, NLP, Algorithms and Soft Computing.

Besides having research interest in Computer Science also did a measurable amount of research in the field of TQM, BPR and HRM. She has supervised several projects for UG and PG courses and authored and co-authored several publications in journals, Books, and conferences including Research Articles, Chapters, Conference papers, Conference proceedings and Edited Books.

Dr. Amit Sharma (Ph.D.) is Associate Professor in the School of Computer Applications at Career Point University, Kota (Raj.). He received his M.Tech and Ph.D. in Computer Science & Engineering. He has over 15 years' teaching experience of higher education, including Under Graduate and Post Graduate courses. He is also Supervised many scholars of M.Tech and Ph.D. He also works as an external examiner in various universities for PhD evaluation. He has published 7 patents on the latest technologies in computer science. He has also published 30+ Research Papers in various International conferences, UGC journals. Scopus Index, and 8 National journal Research papers. He has also published a Book for B.Tech Mechanical Engineering on CAD Subject. He is worked as a coordinator for organizing 4 International conferences on multidisciplinary topics. His areas of interest are IoT, Distributed systems, Cloud Computing, Cluster And Grid Computing, Big Data, Deep learning, Data Mining, Machine Learning, Mobile Security, Image Processing, Service Oriented Architecture, Wireless technology, and Computer Networking.

Contributors

Naved Ahmad
Jamia Hamdard University
Delhi

Usha Badhera
Jaipuria Institute of Management
Jaipur

Rajiv Gill
Department of CSE
Uttaranchal University
Dehradun

Ganapati Hegde
Dayanand Sagar Institutions
Bangalore, India

Hithesh R.
Dayanand Sagar Institutions
Bangalore, India

Abid Hussain
School of Computer Applications
Career Point University
Kota (Raj.)

Anubha Jain
Department of CS& IT
IIS (Deemed to be) University
Jaipur (Raj.)

Ayush Kumar Jha
Jamia Hamdard University
Delhi

Kapil Joshi
Department of CSE
Uttaranchal University
Dehradun

T. Kohila Kanagalakshmi
Dayanand Sagar Institutions
Bangalore, India

Ihtiram Raza Khan
Jamia Hamdard University
Delhi

Krishananjali Magade
School of Engineering and Technology
Career Point University
Kota (Raj.)

Minakshi Memoria
Department of CSE
Uttaranchal University
Dehradun

Pooja Nahar
S.S. Jain Subodh PG College
Jaipur

Anjali Naudiyal
Department of CSE
Uttaranchal University
Dehradun

C. Priya
Department of Computer Applications
Dr. M.G.R Educational and Research Institute
Maduravoyal, Chennai, Tamilnadu, Chennai

Himanshu Rawat
Jamia Hamdard University
Delhi

Amit Sharma
School of Computer Applications
Career Point University
Kota (Raj.)

Amita Sharma
Department of CS & IT
IIS (Deemed to be) University
Jaipur (Raj.)

Navneet Sharma
Department of CS& IT
IIS (Deemed to be) University
Jaipur (Raj.)

Ajay Singh
Department of Research
Uttaranchal University
Dehradun

Srivatsala V.
Dayanand Sagar Institutions
Bangalore, India

Suneetha V.
Department of Computer Applications
Dayanand Sagar Institutions
Bangalore, India

Salini Suresh
Department of Computer Applications
Dayanand Sagar Institutions
Bangalore, India

Garima Tyagi
School of Computer Applications
Career Point University
Kota (Raj.)

Priyanka Verma
Department of CS& IT
IIS (Deemed to be) University
Jaipur (Raj.)

Ayush Kr. Yogi
School of Computer Applications
Career Point University
Kota (Raj.)

Chapter 1

Intelligent Application Development in Cloud with IoT

Ayush Kr. Yogi and Abid Hussain

1.1 INTRODUCTION

Cloud computing represents a boost in network use in respect to providing and accessing services over the internet. Before taking a dive into the intelligent application developments in cloud computing with IoT, the real meaning of all the related key terms needs to be explored. What is the cloud? Simply, we explain the cloud as a network of interconnected devices, computing machines, resources, databases, and more working in a synchronous and/or asynchronous manner to perform and provide results based on requested queries and services that a particular cloud offers.

Intelligent means capable of taking decisions on real-time problems by using available resources and providing an optimal solution. Concerning the cloud, we say the application or device is intelligent if it produces result by taking parameter data from a run-time process, applying suitable algorithms (intellectual decisions to select appropriate method/algorithm), and producing output. In its simplest term, IoT (Internet of Things) are devices that are connected to a network and use the internet to operate. Intelligent application development in cloud computing with IoT involves creating applications for IoT devices that have intelligence enabled, such as artificial intelligence (AI), or algorithms that use an artificial neural network (ANN) paradigm for learning and producing results.

1.2 CLOUD COMPUTING

Traditional computing differs from cloud computing in that the latter supports the facility to compute over the network rather than buying or paying for individual software, infrastructure, or platform. This revolution in computing has reduced cost and increase the efficiency about the cloud services for the cloud subscribers. The cloud is like sliced fruit, in which slices of platform, infrastructure, and software are packed with different service-oriented features.

DOI: 10.1201/9781003406723-1

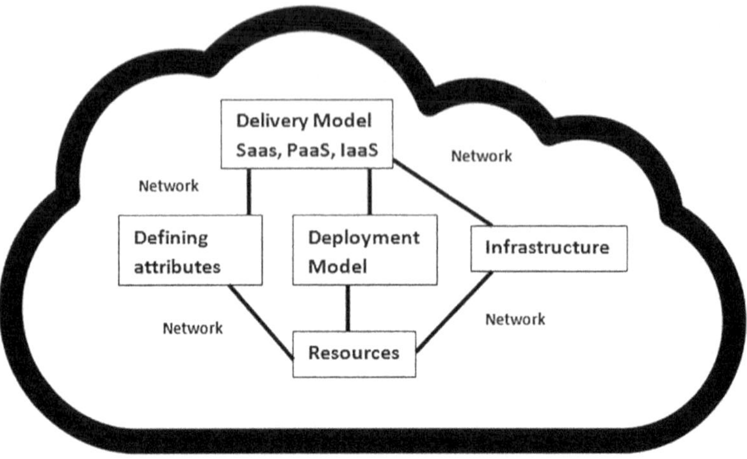

Figure 1.1 Cloud configurations.

The cloud is not a technology but a model in which the cloud is the pool of resources including server, software, storage, utility applications, infrastructure, and platforms. The cloud computing model that was most appreciated and accepted was from the National Institute of Standards and Technology (NIST) that presented the document in 2011 titled "NIST Cloud Computing Reference Architecture".

1.2.1 Cloud Deployment Model

Deployment means the arrangement of services according to the requirements of the customer, user, or organization. This model defines the boundary conditions of cloud accessibility. The NIST defines four common deployment models as follows.

1.2.1.1 Public Cloud

The public cloud refers to services that are accessible to all subscribers without any limitations. Based on its physical location, the public cloud is also called an external cloud. Because the public cloud can be accessed by any subscriber, it has a large number of services, such as resource sharing, network, data storage, software or application development platform, and many more. The public cloud is managed and controlled by computing vendors.

Some popular public cloud vendors are Amazon Web Services (AWS), Google Cloud, Microsoft Azure, and Salesforce. The public cloud provides a one-to-many relationship to its customers.

The most promising advantages of the public cloud include the following:

- Promotes multi-tenancy at its highest degree
- Provides a large magnitude of operations

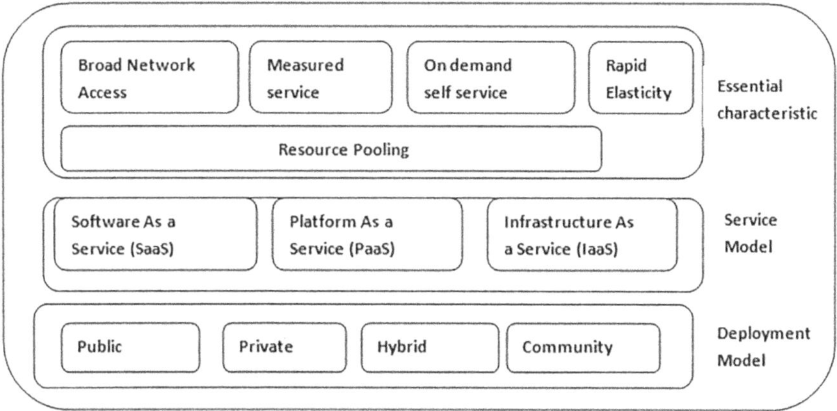

Figure 1.2 NIST cloud computing model.

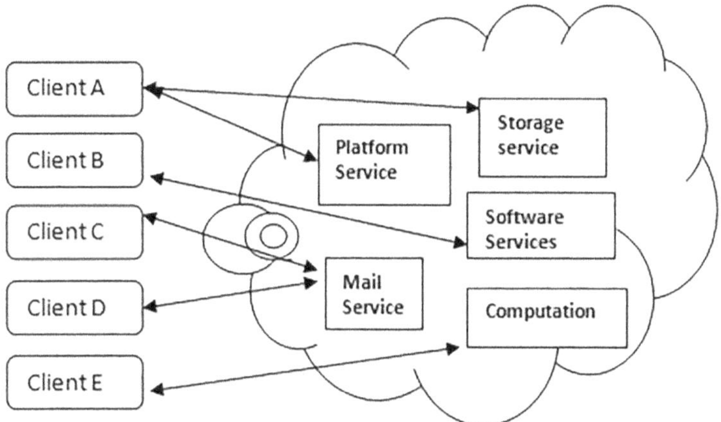

Figure 1.3 Public cloud.

- Makes state-of-the-art technology affordable
- Ensures better quality of services
- Offers customers superior service at lower cost

1.2.1.2 Private Cloud

The model of cloud where access of services are bound, based on some limitations such as physical location or privacy policies, is referred to as "private cloud". Based on its physical location, it is also referred to as "internal cloud". Categories of internal cloud are on-premises and off-premises. Different services are delivered based on storage, server optimization, email server, and the resource provided. Organizations have their own cloud to

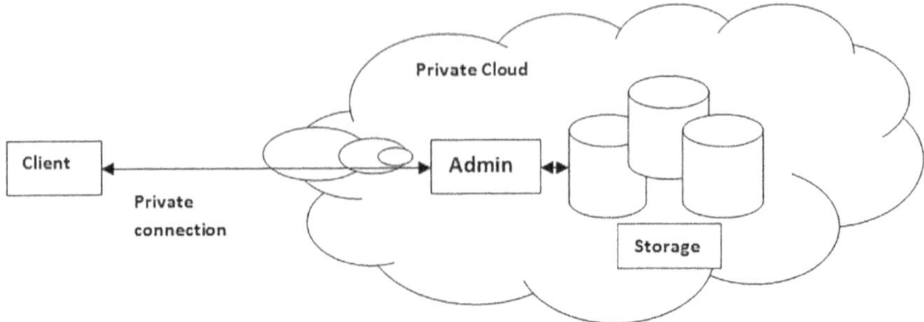

Figure 1.4 Private cloud of storage service provider.

provide services to their location-dispersed institutes and do not require an external entity to connect within the community.

1.2.1.2.1 On-Premises Internal Cloud

A cloud that resides physically and within a consumer's network boundary to provide services based on its configuration is known as an "on-premises internal cloud".

1.2.1.2.2 Off-Premises Internal Cloud

If the cloud does not exists within consumer's own network boundary but can be accessed and managed by a consumer's organization, it is referred to as an "off-premises internal cloud".

Private clouds may or may not provide a large number of services because the private cloud vendor has a configuration such as a platform, infrastructure, software, other utility, or resource services based on particular target customers or organizations. The advantages of private cloud are as follows:

- Has a one-to-one service relationship between cloud vendor and consumer
- Provides a more secure and private computing environment
- Resource dedication is high for consumers
- Consumers can hold control over the environment

1.2.1.3 Community Cloud

This model provides its services to a particular community of members. In other words, community cloud offers a single or selected group of services targeting organizations with the same interest. Community cloud is not limited to customers who are on premises; they may also be off premises. Community cloud vendors may provide services for organizations with the common interest, for example, of digital marketing. The community cloud provides service to all the organizations on a single cloud model of that vendor.

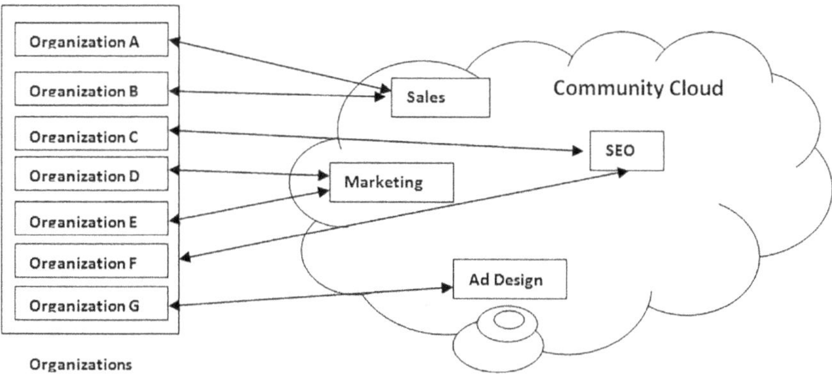

Figure 1.5 Community cloud.

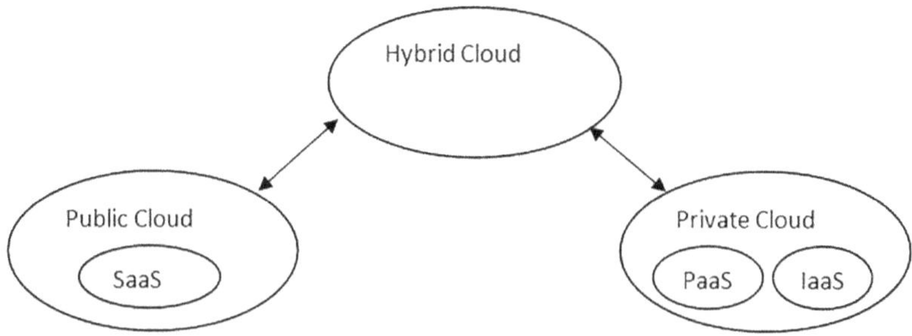

Figure 1.6 Hybrid cloud.

The advantages that make popular this model are as follows:

- Multi-tenancy is preserved as public cloud
- Degree of security is provided at the level of private cloud
- Pay-per-use billing for consumers, which is cost effective compared to a private cloud

1.2.1.4 Hybrid Cloud

This cloud model generally inherits public, private, and community cloud model's on-premises and off-premises features, but with limitations. Some functionalities are provided in a public manner and some functions or services are provided in a private cloud–based model.

A hybrid cloud offers sharing and storage of data with an on-premise private cloud, an off-premise community cloud to share the platform, a public cloud to access the membership of the cloud, and so on.

Thus, making cloud computing model progressive and intelligent entails an evolutionary process that enhances its architecture and service model to offer dynamic, robust, and flexible services to its customers.

The advantages that make this model popular are as follows:

- Flexibility
- Unlimited storage
- Cost efficiency
- Scalability

1.2.2 Cloud Service Model

A cloud service layer model offers various services tailored to the consumer's needs, encompassing infrastructure as a service (IaaS), platform as a service (PaaS), and software as a service (SaaS).

This model has a structure based on providing these services using the internet or a network. The user, customer, or client has the benefit of on-demand service at any time. Scalability is high enough so that the user can choose an appropriate vendor and access the service by subscribing or registering as a client. The following sections discuss each service section.

1.2.2.1 Infrastructure as a Service (IaaS)

Infrastructure such as processor, memory, storage, and network resources is provided over the internet as a cloud. All the computing is performed in

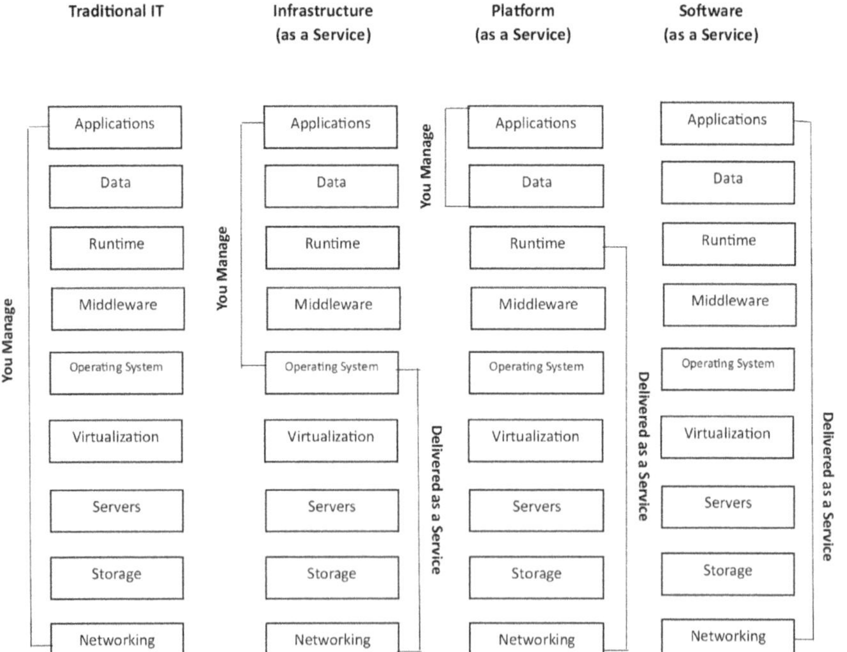

Figure 1.7 Cloud service model.

the cloud (the hub of services). In reality, calculations are performed at the machine, not over the internet, and then the network or internet is used to transmit the data or processing commands/instructions. A single piece of hardware is simulated through different instances, thereby introducing the concept of virtualization of resources or infrastructure components, such as virtual processors and multiple threads of server services, which are sometimes termed server instances. In other words, IaaS can also be referred to as hardware as a service (HaaS).

Examples of infrastructure that offer virtualization instances of HaaS include Amazon EC2 and Google Compute Engine.

1.2.2.2 Platform as a service (PaaS)

What does platform mean? A platform has the synchronized environment of hardware, operating system, middleware (applications), and runtime libraries to provide a full development and deployment environment. The advantage of this service for the customer is the ability to avoid scaling up investments for each required phenomenon that requires investment. Therefore, if a client needs to develop an Android app, then he/she does not need to install Android in its own platform—a cloud-offering service vendor can provide this facility to the customer. The available vendors that offer such cloud services are Google Cloud App Engine, Microsoft Azure Platform, Salesforce, and GoGrid Cloud Center. There is vendor lock-in problem associated with PaaS—apps developed over one PaaS vendor platform cannot be migrated to another PaaS vendor platform. Flexibility is needed regarding migrating one vendor PaaS to a second one.

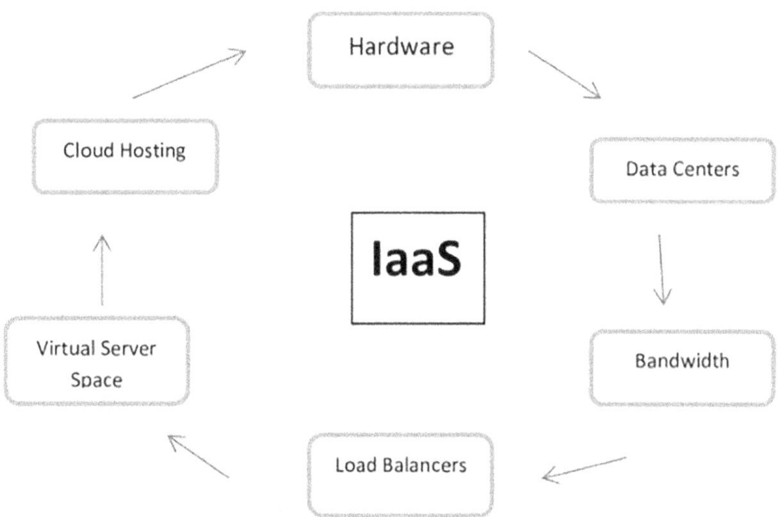

Figure 1.8 Infrastructure as a service.

Figure 1.9 Software as a service.

1.2.2.3 Software as a Service (SaaS)

In traditional computing, a user needs to install the software or application that he/she wants to use. In the SaaS cloud service model, however, the user does not have to install the application or software to use it.

SaaS is designed above the PaaS layer. The entire hardware or platform configuration (software/application in cloud) required is provided by the PaaS layer. Based on the load or access of the application, multiple threads are created to ensure that data integrity is not compromised. To access the application or software over the cloud, the solution lies in using a thin client interface via the internet.

E-mail facility, customer relationship management (CRM) package of Salesforce, SAP of ERP, Google apps, and Microsoft Office 365 are popular "on demand" SaaS offerings.

1.3 IoT(INTERNET OF THINGS)

In the current working, entertainment, and business scenarios, the internet plays a pivotal role in achieving success. As internet usage becomes more cost-effective, its popularity grows due to its versatility.

Advancement in technology and network development have increased the use of the internet in this global era. Efficient and effective request-response phenomena have accelerated the use of the internet in the cyber world.

1.4 INTELLIGENT APPLICATION DEVELOPMENT WITH IoT IN CLOUD COMPUTING

Efforts are continuously being made to implement intelligence in the cloud model. The proposed intelligent cloud computing architecture includes a behavior control mechanism that adjusts based on the desired analytical operations, planning, and execution of changes, thereby altering the

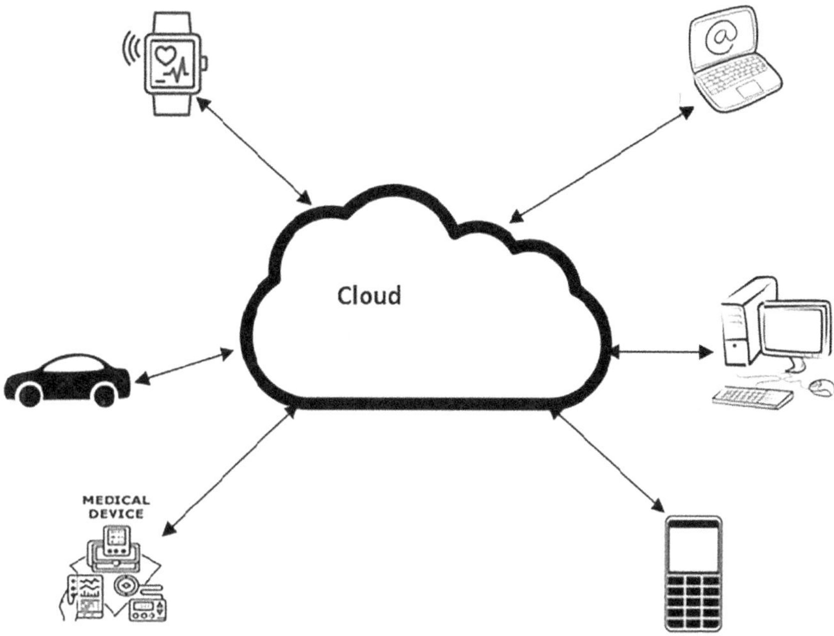

Figure 1.10 Cloud-enabled IoT environment.

working style from a request-response approach. A review on the intelligent application development in cloud and future aspect is a vast with distributed nature. As we talk about intelligent applications in cloud, there are different and wide streams. Based on cloud configuration and the services provided, applications development has variety. Variety of intelligent applications with IoT devices in cloud has proposed and processed in market regarding to health care, agriculture, financial data sets, big data in commercial organizations, sensor technology, customer relations etc.

1.4.1 Agritech

Agriculture is the most important economic sector of every country, requiring the most technical development. As a progressive approach and developments in internet-based technology, AI, cloud computing, and IoT, efforts to develop intelligent applications are being implemented successfully. Because agriculture has variety of activities and climate conditions as greenhouse operations, lightening condition, irrigation, fertilization, pest control, and air humidity, a variety of intelligent applications has been developed. As per the current scenario, the global market of agriculture IoT products is expected to reach nearly one billion US dollars by 2027. With the advancement in sensor technology and cloud computing with IoT devices, intelligent

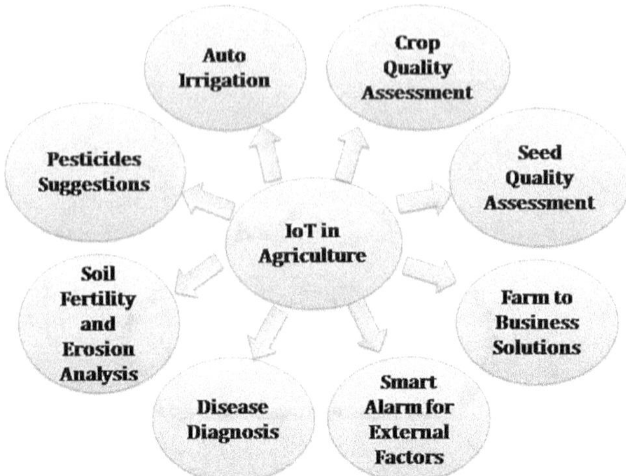

Figure 1.11 IoT for agriculture.

applications have developed for agriculture. We summarize five major intelligent applications developed in IoT for agriculture.

"Farmapp" is an integrated pest management software that captures satellite images of agricultural land and, using IoT sensors, collects data regarding soil fertility, water percentage in soil, pesticide ratio, humidity, temperature, and so on. For various agricultural activities, an intelligent cloud-based algorithm is applied to the collected data that produces an analytical result.

For livestock monitoring, smart IoT sensors are used to monitor the herd.

Cowlar, an intelligent IoT system using cloud computing, is used to monitor cattle reproduction and optimize milking and farm performances.

Sense Fly, a modern drone that performs intelligent functions for agricultural activity such as plant counts, determination of the chemical composition of soil and plants, irrigation, fertilizer uses, the presence of pests, and more. This reduces the farmer's time and labor and speeds up by 40–60 times the process of gathering such information.

Other intelligent IoT applications based on cloud computing in agriculture are Mothive, Arable, and IoT agriculture platforms such as Semios that provide full functional prediction from crop management to soil fertility.

Technology should be beneficial to the economic, health, and living conditions for agriculture-involved people. Spraying pesticides is an important activity in agriculture to protect growing plants from insects, but it also has negative chronic health impacts that can lead to diabetes, hypertension, ophthalmic disorders, asthma, and dry skin.

Drone systems with enabled intelligent optimized algorithms using cloud computing can be operated and controlled remotely, instead of manually, with all the advantages of economic and health. Some of the big advantages for the farming community in using IoT sensor–enabled drones includes time and money savings and health benefits for those involved in the spraying of pesticides.

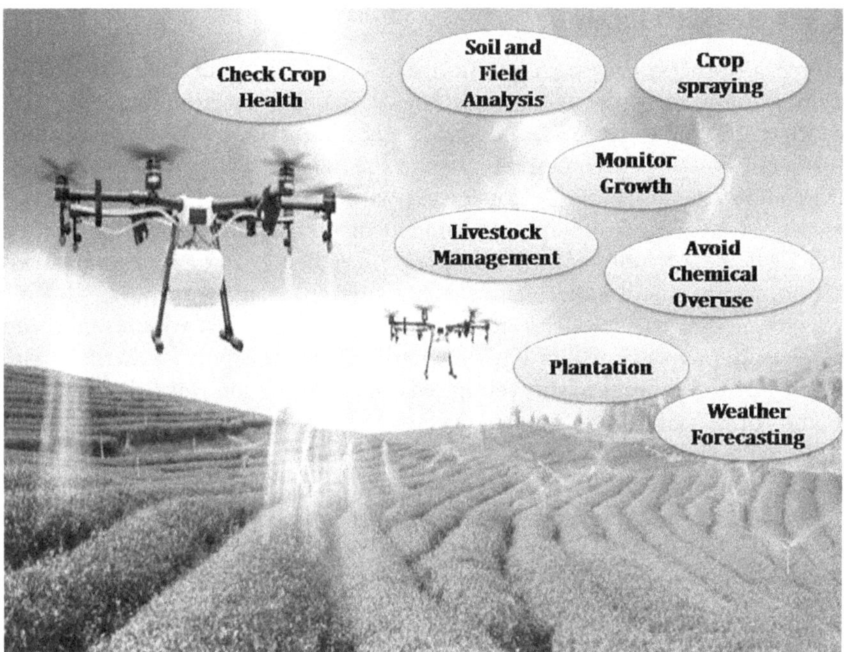

Figure 1.12 Drone use for agriculture.

1.4.2 Intelligent Medical IoT Applications

Technology that enables fast medical services has become feasible through IoT devices leveraging cloud computing. This sector holds vast potential, with numerous IoT devices operating in the modern era. From testing or diagnosis to report generation and the transmission of live data over the cloud for generating intelligent solutions, IoT combined with cloud computing consistently boosts this sector. Various segments of IoT devices with intelligent application implementations, such as neurological devices, anesthesia machines, respiratory devices, and more, contribute to this advancement.

Hip replacement in any person incorporates a concealed sensor that detects the surgery site's condition for recovery. It periodically transmits data according to preset time parameters and sends the collected data to the control system via the cloud for monitoring purposes.

Persona IQ is the intelligent sensor-based IoT application that senses steps, speed, distance, and closed path of motions and knee functions following surgery.

As an external IoT, MY01 uses a continuous compartment syndrome monitoring system for real-time monitoring.

OTOROB is an IoT-based telerobotic consultation system that suggests treatment; doctors use it as an assistant to confirm their decisions.

1.4.3 Smart Home

Home automation uses a collection of smart IoT devices with the cloud computing model to provide technological benefits. Starting with the Bluetooth technique to connect and share operating data, control, and provide results, it has gained wide popularity in the market. Due to its short range, it provides the advantage of operating within a confined area of approximately 10 meters. To obtain and implement the advantages of cloud-based IoT devices, a remote controlled operation is needed. For such implementation, the protocol called message queuing telemetry transport (MQTT) was developed to establish remote connection via Wi-Fi, which sends and receives data from sensors implemented on the transport network. Raspberry Pie, for example, is a web-based user-friendly interface that controls home appliances.

To make home appliances smart, a module must be synchronized to make IoT devices controlled remotely. Some major components are Node-MCU (ESP32), Amazon Alexa, Sinric, and Arduino software (IDE).

Amazon Alexa is a speech recognition system that accepts voice inputs and converts them to commands, sends them over the internet to the authenticated website or portal, and produces output with the home's connected devices.

1.4.4 IoT in Industrial Operation Control

In industries, different operations need to be automated. Depending on the type of industry, such operations include manufacturing, sales and marketing, row material management, supply chain management, packaging, accounting, and employee attendance and record management. For various industries, this technology migrates to different units. In the production process, sensor technology facilitates the scanning of packed items. If any defects are detected in the packed product, it discards that item from the conveyer belt by using an applied mechanical belt mechanism. Signaling to the specified system is automated through network communication to the control system. Additionally, in food production units, indoor air quality is measured.

Different sites of the same organization connected via the network have a central controlling system that displays units showing the air quality of food production units. IoT sensors may be connected to a private cloud that analyzes the data and controls the operation on site.

1.4.5 Smart Switchboard

The switchboard is connected directly to the electricity. An idea has been proposed to design a prototype model for children's safety using an ESP module and AWS. This IoT switchboard operates on a paradigm in which the ESP module senses the load and sends it using Wi-Fi to the receiving module. The idea proposed is not real-time implementation with electric circuitry but provides an IoT-based approach that can monitor the activity of a child near an electric panel and switchboard contact.

1.4.6 Lifesaver Intelligent IoT Tools

As technology continues to advance, progress in life processes is also reflected. Intelligent IoT applications and tools have been developed to protect people from accidents. Some of them are as follows.

- To provide road safety, an intelligent IoT–based system named Bumblebee has been introduced. The system prevents accidents by notifying a central control system of heavy traffic conditions and aggressive driving behaviors, giving emergency vehicles a clear pathway, and alerting nearby cars to maintain a safe distance from each other.
- To assist in smart city initiatives, smart waste and recycling management is implemented. BigBelly is a platform that informs a central authority control system when a waste collection bin is full or has reached its storage limit.
- Air Quality Egg is an air pollution monitoring app that notifies the air quality control system of pollution levels and indicates alarming situations based on the pollution and air quality levels.
- Forest fires are natural hazards that sometimes spiral out of control. If detected early, appropriate preventive measures can be taken to contain them. Insight Robotics is an IoT-based device that detects forest fires and notifies a control system via the network.
- In patient medical care, Proteus Discover is a small IoT system comprising a mobile phone application and an ingestible sensor. It assists medical specialists in gaining comprehensive insight into their patients.
- Vigo Smart Headset is an IoT tool used to prevent accidents due to driver fatigue.

1.5 FUTURE TRENDS IN DEVELOPMENT OF INTELLIGENT IoT APPLICATIONS USING THE CLOUD COMPUTING MODEL

Converting IoT applications into intelligent IoT-cloud applications is the right approach for future trends. As discussed previously, numerous intelligent applications have been developed, and there is a continuous stream of developments toward intelligent applications in cloud computing with IoT, suggesting limitless future potential. Some of the streamlined synchronous aspects for the future scope are highlighted in the following sections.

1.5.1 Vendor Collaboration

The technology is for people, not for personal gain. This theme increases the popularity of developed technology. As the cloud computing model has the disadvantage of vendor lock-in, there is a future possibility of vendor mashup, wherein different vendors will adopt a collaborative business environment. In the vendor mashup model, a collaborative platform of cloud computing can be established. Because application developed using one vendor's platform cannot run on other vendor's platform, user migration is locked.

1.5.2 Intelligent Cloud

Cloud computing is a model with the potential to become intelligent, which would offer a revolutionary update in the same field. Utilizing AI and machine learning techniques and algorithms in cloud configuration may reduce the downtime and maintenance costs and time that subscribers currently endure. Implementing artificial intelligence through the automatic scaling of services and resources with auto update components is possible, but the cost is likely to be high.

1.5.3 More Security

In a public cloud, users do not have control over the data, opening up the possibility of improving security concerns. Cloud computing with IoT devices has also shown various vulnerabilities such as distributed denial of services (DDOS), data leaks on social media, cyber threats, and malware, revealing the need to improve security in the cloud paradigm.

1.5.4 Edge Computing

Dependency on the cloud and the growing collaborative approach necessitate refinement to scale up computing at the edge of need. Edge computing has the potential to redefine the computing of resources and energy efficiency in parallel to cloud computing. The future holds great promise for producing such a cloud computing model. With the increasing applications of IoT and smart apps, edge computing will offer efficient and cost-effective solutions to its subscribed customers.

1.5.5 Quantum Computing

Quantum computers store data in qubits, which accelerates processing. Utilizing quantum computing to process cloud services, such as SaaS, can significantly enhance computing speed and software development over the cloud. Additionally, storing data and paying for storage over the cloud will reduce subscription fees for large organizations. Increased storage capacity with efficient processing power will enhance the efficiency of the cloud computing environment. Moreover, selecting PaaS services with the incorporation of quantum computing will offer efficient choices for subscribers.

1.5.6 Energy Efficient

The energy consumption of cloud-based data centers is high compared to an office running with different computer systems. Reducing energy consumption is a necessity, and there should be techniques available to optimize energy use in cloud data storage centers. Additionally, these centers emit heat that must be managed through cooling systems. Significant efforts are required to address and solve these problems effectively.

1.5.7 Cloud Disaster Recovery

Cloud-based recovery is available for organizations that subscribe to a vendor's cloud. Organizations receive service over the cloud to manage their most critical systems, such as their centralized data center. When an organization's central data center crashes, the cloud restores the data at the state of its critical disaster. In the future, IoT-enabled accessibility should be preserved.

1.6 CONCLUSION

With the rapid and increasing adoption of cloud computing by medium to large organizations in accordance with the service required, client organizations are reaping the benefits of lower capital and operational expenditures. To manage resources efficiently in terms of both technology and finances, client organizations have transitioned to focusing on the logical implementation of techniques, leaving the burden of resource implementation and maintenance to the cloud.

According to the current market, the market size of global cloud computing was valued at USD 483.98 billion in 2022 and is expected to grow at a compound annual growth rate of 14.1% from 2023 to 2030. The SaaS segment extended its business to 50% in 2022, and the private deployment segment has nearly reached that level at the limit of 40% in 2022. After Covid-19, the demand of cloud computing has increased exponentially.

With the ongoing surge in demand for cloud computing, the future holds significant financial investment opportunities and the potential for updating the cloud model. The integration of IoT with cloud computing, incorporating AI and machine learning, is underway, leading to the development of intelligent applications.

Chapter 2

Significant role of IoT in Cyber-Physical Systems, Context Awareness, and Ambient Intelligence

Krishananjali Magade and Amit Sharma

2.1 OVERVIEW OF INTERNET OF THINGS

The Internet of Things (IoT) is a network of gadgets that transmit data to a platform to allow for automated control and communication. IoT links humans or machines to other machines. It mainly links the actual hardware to the digital interfaces.

These actual objects contain sensors that transmit information to a central platform. They frequently get directives based on this information. The platform also does data analysis to provide customer customization and insights to company owners.

Smartwatches, injectable ID chips for animals, temperature sensors for jet engines, and voice controls for the house are a few examples of these gadgets. Amazon Echo and Google Home are two prominent examples. A vending machine at Carnegie Mellon that was linked to the internet is believed to be the first basic IoT gadget.

According to IoT Analytics' State of IoT Summer 2021 report, around 27 billion IoT connections will be operational by 2025. The technological advancements that have aided in IoT taking off are:

1. RFID tags: RFID tags are wirelessly communicative, low-power semiconductors. They may be incorporated into bigger, more expensive pieces of machinery since they are compact and accessible.
2. Wireless and cellular networks: As wireless and cellular networks are more widely available, it is simple for devices to maintain connectivity when travelling. With the help of Wi-Fi 6 and 5G networking advancements, IoT networks now have the high bandwidth they need to maintain their low latency rates.
3. IPv6: When IPv4 was replaced by IPv6, enough IP addresses were generated to support all devices worldwide for the foreseeable future.
4. Cheap, low-energy sensors: The semiconductor industry has expanded rapidly and is still expanding. Today's inexpensive sensors don't use a lot of electricity. The tiniest gadgets can run artificial intelligence (AI) because of their powerful computational capabilities.

 DOI: 10.1201/9781003406723-2

5. Powerful machine learning algorithms: Today's machine learning (ML) algorithms are more sophisticated, complicated, and effective than ever. The new technologies like neural network training currently operate 44 times faster than they did in 2012.
6. Cloud computing: Cloud adoption, which was already on the rise, has been accelerated due to the pandemic. Companies are looking to 50% or more of their applications being on the cloud. Cloud makes it easier for data to be transferred between IoT devices and the AI platforms that analyze them.

A genuinely robust IoT system requires combining all this technology. Other technological aspects such as edge computing may need to be considered based on business requirements.

2.1.1 An IoT System Consists of Four Crucial Components to Make It Work

Physical devices: Physical devices with sensors are the starting point of any IoT device. A camera with motion sensors, a car that gauges tire pressure, and Amazon's Alexa that plays songs on request are all physical devices. They provide raw, base data about operational environments and user inputs.

Connectivity platform: These devices can be hooked onto a network in several ways. The most common are Wi-Fi and cellular networks. 6LoWPAN, or any low-power radio, can connect to the internet thanks to low-power wireless personal area networks. Zigbee is a wireless network that operates largely in industrial environments and uses little power and low data rates. OneM2M standards and others aim to formalize machine-to-machine communication patterns.

Analytics platform (IoT platform): Machine learning plays a significant role in analyzing the massive data sent by physical devices. Algorithms are trained to read this data and provide in-depth analysis backed by historical data. This can also be used to predict future behavior.

The platform has two primary purposes. The first is to facilitate informed, data-driven decisions at the business level. The second is to send automated instructions back to the devices. This allows the IoT devices to respond intelligently to user inputs or environmental changes.

Configuration manager: The sheer number of devices and data involved makes IoT-based systems complex to administer. A configuration manager gives the business a bird's eye view of the devices involved, the various parameters required to monitor and control machines, and the algorithmic tweaks required with new additions. Security patch management must ideally also be a part of this setup.

Dashboard: The dashboard makes sense of all the data mined and analyzed from the IoT system.

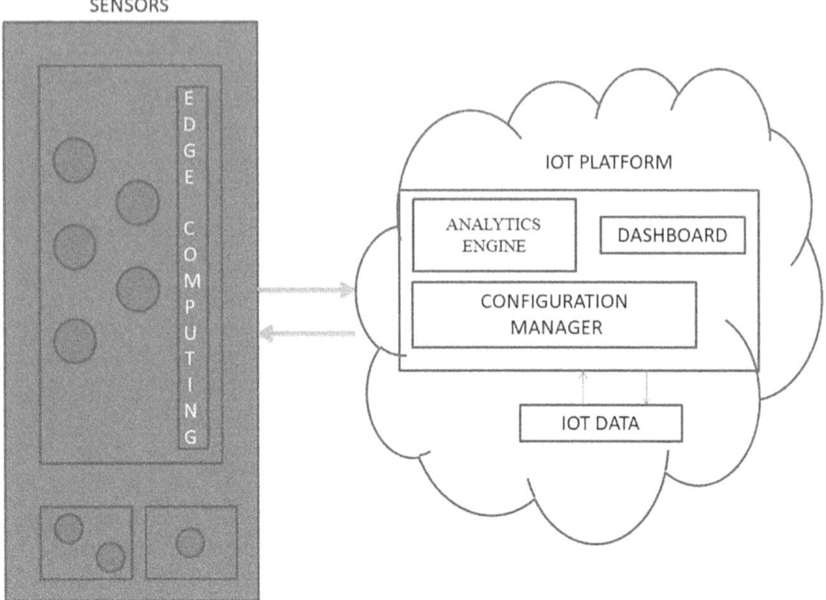

Figure 2.1 Component of an IoT platform.

2.1.2 IoT Device Management

To effectively support, monitor, and sustain the ever-expanding range of interconnected devices in a residential or commercial network, a diverse set of methods, resources, and technologies must be employed. As the number of network-enabled devices continues to rise, the demand for IoT device management software is also increasing. Cisco's Annual Internet Report (2018–2023) predicts that by 2023, there will be 29.3 billion networked devices, equivalent to 3.6 devices per person on Earth. The significance of IoT device management arises from two key factors: pull and push.

Intelligent management of IoT devices serves as the cornerstone for advanced analytics, seamless automation, internal efficiency, and innovative business models, creating a definite pull factor. This is evident in business models such as servitization, which replaces outright equipment sales with equipment leasing and services driven by IoT data. Moreover, there is a push aspect as the adoption of connected devices continues to grow. Without IoT device management, employees may inadvertently introduce additional endpoints to the company network, resulting in a significant increase in shadow IT labor.

Based on a 2020 study analysis by Valuates Reports, the demand for IoT device management is projected to grow at a compound annual growth rate (CAGR) of 22.6% between 2021 and 2026. As a result, the IoT device management market is expected to reach a value of $6.25 billion worldwide by the end of this forecast period.

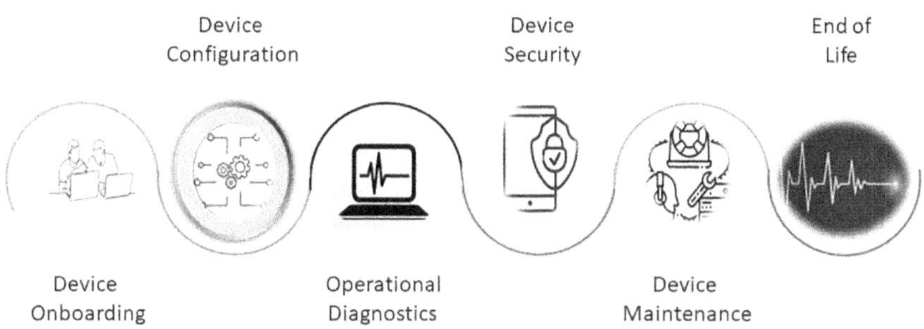

Figure 2.2 Key components of IoT device management.

IoT device management encompasses both the procedures and the tools necessary for controlling an IoT environment, as mentioned previously. Over the lifecycle of an IoT device, several crucial operations are involved, including the following:

Device onboarding: When IoT devices are initially powered on, they must be integrated into the network. However, unlike traditional devices, they lack a comprehensive, standalone interface to facilitate the onboarding process. Device onboarding entails tasks such as verifying credentials, establishing authentication protocols, assigning device identities, and more.

Device configuration: Each IoT device connected to a network needs to be configured to align with a company's specific requirements. For example, if a company manages a fleet of connected vehicles, they may want to group devices based on their usual operating locations or destinations.

Operational diagnostics: Diagnostics provide valuable insights into the performance of IoT operations. Since many IoT devices lack sufficient memory or computational power to analyze diagnostics locally, a centralized IoT device management capability is essential for collecting and interpreting diagnostic data.

Device security: Security plays a critical role in IoT device management. Despite constituting 30% of all endpoints, an alarming 98% of IoT device traffic in the U.S. was transmitted over unencrypted channels in 2020. IoT device administration implements appropriate security measures and brings uncharted endpoints under organizational control.

Device upkeep: In addition to updating device firmware to the latest version, it is crucial to monitor for any security vulnerabilities introduced by recent releases. Similar to onboarding and configuration, IoT device management facilitates over-the-air (OTA) updates in bulk for efficient device maintenance.

End of life: IoT devices that are no longer in use but remain connected to the company network pose significant security risks, as external entities could clandestinely extract data from these devices. Additionally, outdated or malfunctioning gadgets can severely impact operations. End-of-life policies and protocols define the necessary actions for decommissioning an IoT device, environmentally friendly material recycling to reduce carbon footprint, and proper device retirement procedures.

2.2 IoT AND CYBER-PHYSICAL SYSTEMS: INTEGRATION, ARCHITECTURE, AND DESIGN CONSIDERATIONS

Integration, architecture, and design considerations are crucial aspects when implementing IoT and cyber-physical systems (CPS). The successful integration of these systems requires careful planning, collaboration, and adherence to certain design principles.

One of the key considerations in integrating IoT and CPS is the interoperability of devices and systems. Since these systems comprise various devices, sensors, and platforms from different vendors, ensuring seamless communication and data exchange between them is essential. Standardization of protocols, data formats, and interfaces becomes vital to enable interoperability and facilitate the integration process. Another important aspect is the architecture of the combined IoT and CPS infrastructure. A well-designed architecture should accommodate scalability, flexibility, and adaptability. It should be able to handle the increasing number of connected devices, the growing volume of data generated, and the evolving needs of the system. A distributed architecture that can support edge computing capabilities is often preferred to reduce latency and ensure real-time processing.

Security is a critical concern in IoT and CPS deployments. The integration of these systems introduces new attack vectors and vulnerabilities, making robust security measures imperative. Designing a secure architecture involves implementing strong access controls, encryption techniques, authentication mechanisms, and secure communication protocols. Regular security assessments, threat modeling, and continuous monitoring should also be part of the design considerations. Considering the resource constraints of IoT devices, energy efficiency is a crucial aspect. Designing energy-efficient systems and optimizing power consumption becomes essential to extend the battery life of devices and ensure their seamless operation. Techniques such as low-power communication protocols, energy harvesting, and intelligent power management should be incorporated into the system design.

Furthermore, data management and analytics play a vital role in extracting meaningful insights from the vast amount of data generated by IoT and CPS. Designing a scalable and robust data management framework that can handle data collection, storage, processing, and analysis is crucial. Employing advanced analytics techniques such as machine learning and artificial intelligence can enable organizations to derive valuable insights and support intelligent decision-making. Considering the human aspect, user experience (UX) design should not be overlooked. IoT and CPS should be intuitive, be user-friendly, and provide meaningful feedback to users. Considering the context of use and incorporating human-centered design principles can enhance usability and user satisfaction.

Lastly, collaboration and interdisciplinary approaches are essential for successful integration and design of IoT and CPS. Engaging experts from

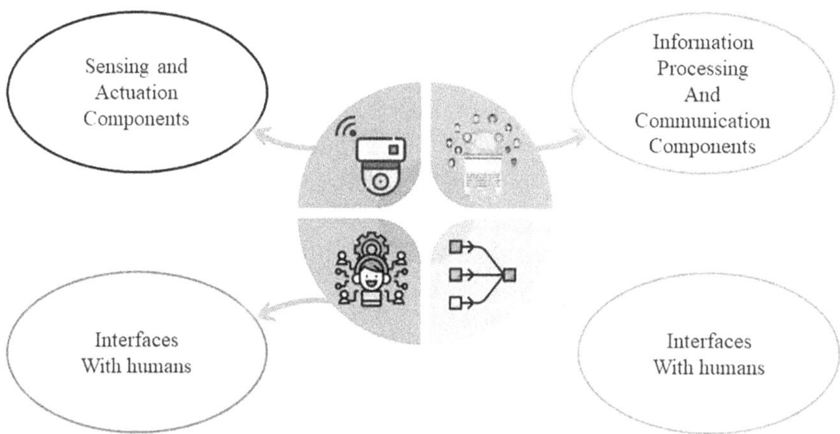

Figure 2.3 IoT in CPS.

various domains such as engineering, computer science, cybersecurity, and data analytics can ensure comprehensive and well-rounded system designs. In conclusion, integrating IoT and CPS requires careful consideration of various factors such as interoperability, architecture, security, energy efficiency, data management, user experience, and collaboration. By addressing these design considerations, organizations can successfully implement IoT and CPS that deliver enhanced automation, efficiency, and intelligence while ensuring security, scalability, and usability.

2.3 CONTEXT AWARENESS IN IoT-ENABLED CYBER-PHYSICAL SYSTEMS: CHALLENGES AND OPPORTUNITIES

Context awareness in IoT-enabled CPS presents both challenges and opportunities. Context awareness refers to the ability of systems to understand and respond to the surrounding environment and user context. In the context of IoT and CPS, this involves capturing and interpreting real-time data from various sensors, devices, and contextual cues to provide intelligent and adaptive functionality. One of the significant challenges in achieving context awareness is the sheer complexity and heterogeneity of the data sources involved. IoT devices generate vast amounts of data with diverse formats, protocols, and levels of reliability. Integrating and harmonizing this data to derive meaningful context can be a daunting task. Additionally, ensuring data quality, accuracy, and timeliness is crucial for reliable context-aware systems. Another challenge is the dynamic and ever-changing nature of the context. Contextual factors such as location, time, user preferences, and environmental conditions are subject to constant variation. Adapting to these changes and updating the context model in real time requires efficient algorithms, scalable architectures, and robust data processing capabilities.

Furthermore, privacy and security concerns arise when dealing with context-aware systems. The collection, storage, and analysis of personal data for context awareness purposes raise ethical and legal considerations. Safeguarding sensitive information, ensuring data privacy, and implementing secure communication channels become critical challenges that need to be addressed. However, context awareness in IoT-enabled CPS also offers significant opportunities. By leveraging contextual information, systems can deliver personalized and adaptive experiences to users. Context-aware applications can optimize resource utilization, automate routine tasks, and provide intelligent recommendations based on user preferences and situational context. For instance, smart homes can adjust lighting and temperature settings based on occupancy and environmental conditions, resulting in energy efficiency and improved comfort.

Furthermore, context awareness enables proactive decision-making and predictive analytics. By analyzing historical context data, patterns and correlations can be identified, allowing for predictive maintenance, anomaly detection, and optimized resource allocation. This proactive approach

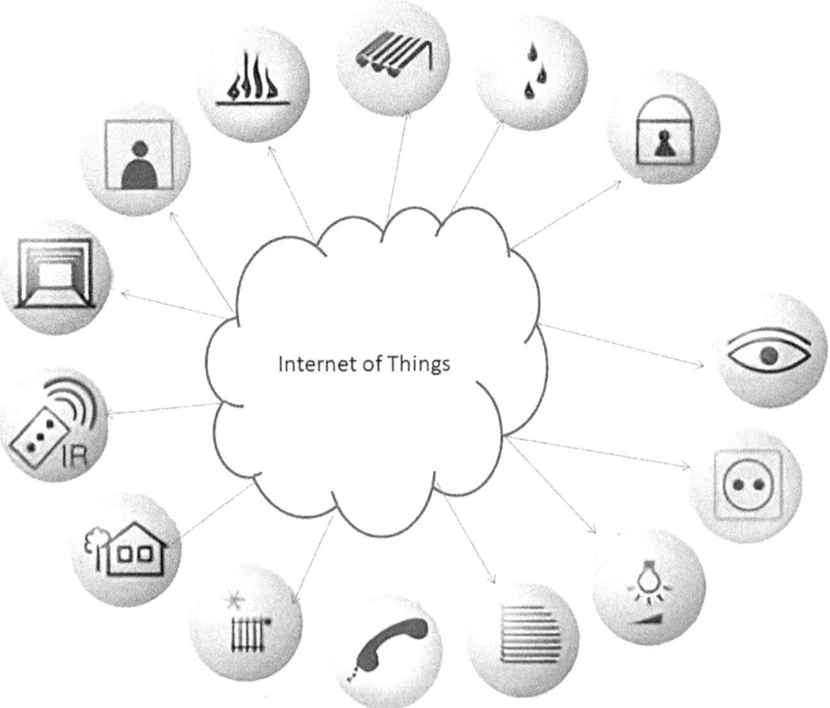

Figure 2.4 Context awareness in IoT-enabled CPS.

minimizes downtime, reduces costs, and enhances system reliability. Additionally, context awareness can enhance safety and security in CPS. Real-time monitoring and analysis of context can enable early detection of potential threats or hazards, triggering timely alerts and appropriate responses. This can be particularly valuable in critical domains such as healthcare, transportation, and emergency management. The achieving context awareness in IoT-enabled CPS presents challenges such as data complexity, dynamic context adaptation, and privacy concerns. However, the opportunities are substantial, including personalized experiences, proactive decision-making, predictive analytics, and improved safety and security. Overcoming these challenges and harnessing the potential of context awareness can lead to the development of intelligent and adaptive IoT-enabled CPS that provides valuable and contextually relevant services to users.

2.4 AMBIENT INTELLIGENCE AND USER EXPERIENCE IN IoT-ENABLED CYBER-PHYSICAL SYSTEMS

In IoT-enabled CPS, ambient intelligence and user experience are essential components. The term "ambient intelligence" describes a system's capacity to detect and react in an unobtrusive and natural way to the presence and demands of people. In the context of IoT, this involves creating an environment that is aware of users' preferences, behaviors, and context, allowing for personalized and adaptive experiences. User experience focuses on designing interfaces and interactions that are intuitive, efficient, and satisfying for users. In IoT-enabled CPS, ambient intelligence enhances UX by creating seamless and contextually relevant interactions. By collecting data from various sensors and devices, the system can adapt to users' preferences, anticipate their needs, and provide personalized services. For example, a smart home system equipped with ambient intelligence can adjust lighting, temperature, and entertainment preferences automatically based on users' habits and current context, enhancing comfort and convenience.

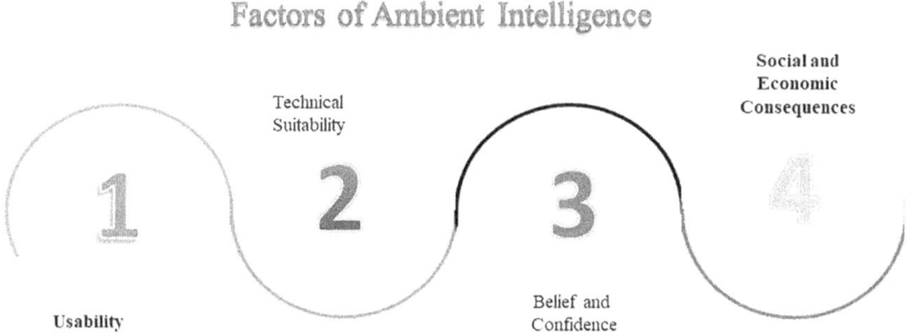

Figure 2.5 Ambient intelligence and user experience in IoT.

2.4.1 IoT-Enabled Cyber-Physical Systems Consist of Two Distinct Components: The Cyber Component and the Physical Component

The cyber component refers to the digital infrastructure and technologies that enable connectivity, data exchange, and computation in IoT-enabled CPS. It encompasses various elements such as communication networks, cloud computing, data storage, software applications, and analytics platforms. The cyber component facilitates the collection, transmission, processing, and analysis of data generated by the physical components of the system. On the other hand, the physical component refers to the tangible and physical entities present in the system. It comprises physical objects, devices, sensors, actuators, and machinery. These physical components interact with the environment and collect data through sensors, perform actions through actuators, and undergo physical processes. The physical component is responsible for sensing and manipulating the physical world, while the cyber component enables communication, control, and intelligence.

The integration of the cyber and physical components in IoT-enabled CPS is what allows for the seamless exchange of information between the digital and physical domains. The cyber component processes the data collected from the physical component, analyzes it, and provides real-time feedback, control, and decision-making capabilities. This connection makes it possible for different applications, including smart homes, smart cities, industrial automation, healthcare monitoring, and transportation systems, to be automated, optimized, and equipped with intelligent functionality. In summary, the cyber component of IoT-enabled CPS refers to the digital infrastructure and technologies that enable connectivity and data processing, while the physical component encompasses the tangible objects and devices that interact with the physical world. The synergy between these two components is what enables the transformative capabilities of IoT-enabled CPS.

2.5 SECURITY AND PRIVACY CHALLENGES IN IoT-ENABLED CYBER-PHYSICAL SYSTEMS

To address security and privacy challenges in IoT-enabled systems, a comprehensive approach is needed. This includes adopting secure design principles, implementing strong authentication and encryption mechanisms, conducting regular security audits and vulnerability assessments, providing user awareness and education, and complying with relevant privacy regulations. By addressing both the cyber and physical components, organizations can enhance the security and privacy of IoT-enabled systems and build trust among users.

In terms of privacy, the cyber component of IoT-enabled systems collects vast amounts of personal and sensitive data from various sources. This

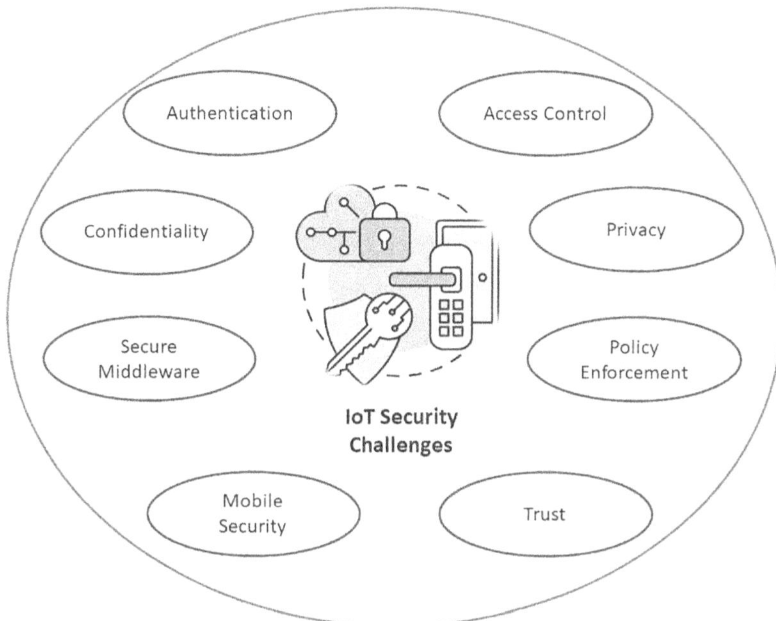

Figure 2.6 Security and privacy challenges in IoT.

data may include personally identifiable information, location data, health records, and behavioral patterns. The challenge lies in adequately protecting this data and addressing privacy concerns. Clear and transparent data handling practices, informed consent mechanisms, and privacy-by-design principles should be implemented to ensure that user data is collected, stored, and processed in a privacy-preserving manner.

On the other hand, the physical component introduces its own security and privacy challenges in IoT-enabled systems. Physical devices and sensors can be physically tampered with or compromised, leading to unauthorized access, unauthorized control, or manipulation of the physical environment. For example, tampering with a smart home security system can lead to unauthorized entry or surveillance. Ensuring physical security measures such as tamper-resistant hardware, secure physical access controls, and device integrity verification becomes important to mitigate these risks.

2.6 REAL-WORLD APPLICATIONS OF IoT-ENABLED CYBER-PHYSICAL SYSTEMS IN SMART CITIES AND INDUSTRIES

The real-world applications of IoT-enabled CPS span across various industries and domains, revolutionizing the way we interact with our environment.

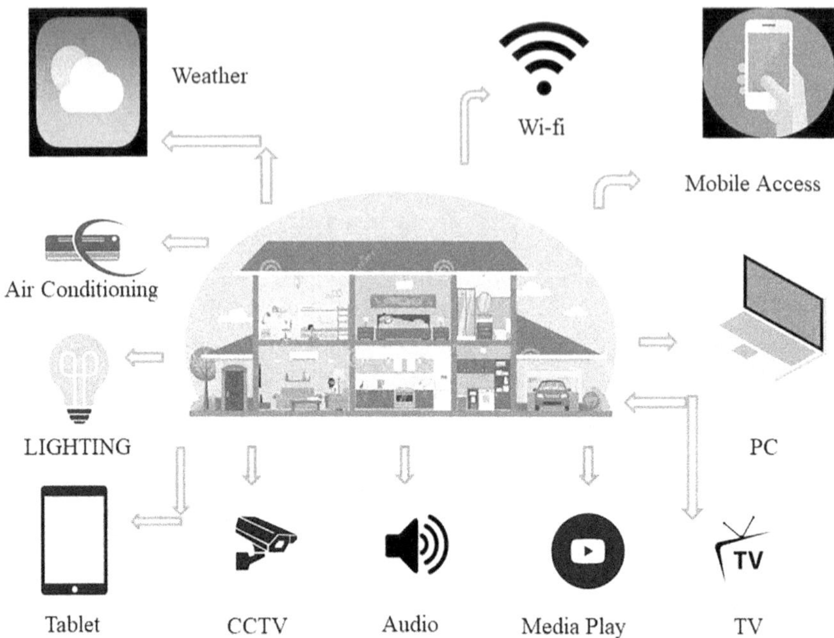

Figure 2.7 Real-world applications of IoT.

In manufacturing, IoT-enabled CPS enables the concept of Industry 4.0, where smart factories leverage IoT sensors, data analytics, and automation to optimize production processes, monitor equipment health, and enable predictive maintenance. This leads to increased efficiency, reduced downtime, and improved productivity. In healthcare, IoT-enabled CPS facilitates remote patient monitoring, wearable devices, and smart healthcare systems, allowing for real-time health data collection, personalized treatments, and proactive healthcare management.

In transportation, IoT-enabled CPS enables smart transportation systems, intelligent traffic management, and connected vehicles, enhancing safety, reducing congestion, and optimizing logistics. Moreover, smart cities leverage IoT-enabled CPS to monitor and manage various urban systems, including energy grids, waste management, public transportation, and emergency services, resulting in improved sustainability, efficiency, and quality of life. These real-world applications of IoT-enabled CPS demonstrate the transformative potential of this technology in diverse domains, paving the way for a more connected and intelligent future.

IoT-enabled CPS has made significant impacts on smart cities and industries, transforming the way they operate and improving various aspects of urban living and industrial processes. In smart cities, IoT-enabled CPS facilitates intelligent infrastructure management, optimizing the use of resources and enhancing the quality of services. For instance, IoT sensors embedded in streetlights can monitor and control lighting levels based on real-time conditions, reducing energy consumption and increasing safety. Smart waste management systems utilize IoT-enabled CPS to optimize garbage collection routes, reduce overflowing bins, and promote efficient recycling. In industries, IoT-enabled CPS revolutionizes manufacturing processes, enabling real-time monitoring of equipment, predictive maintenance, and automation. Sensors integrated into production lines and machinery provide valuable data insights, allowing for optimization of operational efficiency, reduction in downtime, and improved product quality. Additionally, IoT-enabled CPS enables asset tracking and inventory management systems, enhancing supply chain transparency and optimizing logistics. These real-world applications demonstrate the transformative potential of IoT-enabled CPS in creating smarter cities and more efficient industries.

2.7 MACHINE LEARNING TECHNIQUES FOR CONTEXT-AWARE IoT-ENABLED CYBER-PHYSICAL SYSTEMS

Machine learning techniques play a vital role in enabling context awareness in IoT systems. Context awareness involves understanding the surrounding environment, user behavior, and preferences to provide personalized and adaptive experiences. Machine learning algorithms can analyze the vast amount of data generated by IoT sensors and devices to extract meaningful patterns, correlations, and context-related information. These algorithms can be applied to various tasks, such as activity recognition, location estimation, user profiling, and behavior prediction.

For example, machine learning algorithms can learn from historical data to identify patterns in user behavior and predict their preferences or needs in real time. This enables IoT systems to automatically adjust settings, provide personalized recommendations, and anticipate user actions. Reinforcement learning techniques can also be used to optimize decision-making and resource allocation in dynamic and uncertain contexts. Machine learning techniques for context-aware IoT enhance the intelligence and adaptability of these systems, leading to more efficient and personalized user experiences.

Machine learning techniques play a crucial role in enabling context awareness in CPS. Context-enabled CPS involves understanding the context of the system, including environmental conditions, user behavior, and system dynamics, to make intelligent and adaptive decisions. Machine learning algorithms can analyze the vast amount of data generated by sensors and devices in CPS to extract meaningful context information. These algorithms can learn patterns, correlations, and dynamics from historical data and make predictions or decisions based on the current context. For example, machine learning techniques can be used to predict equipment failures or anomalies in industrial CPS, enabling proactive maintenance and

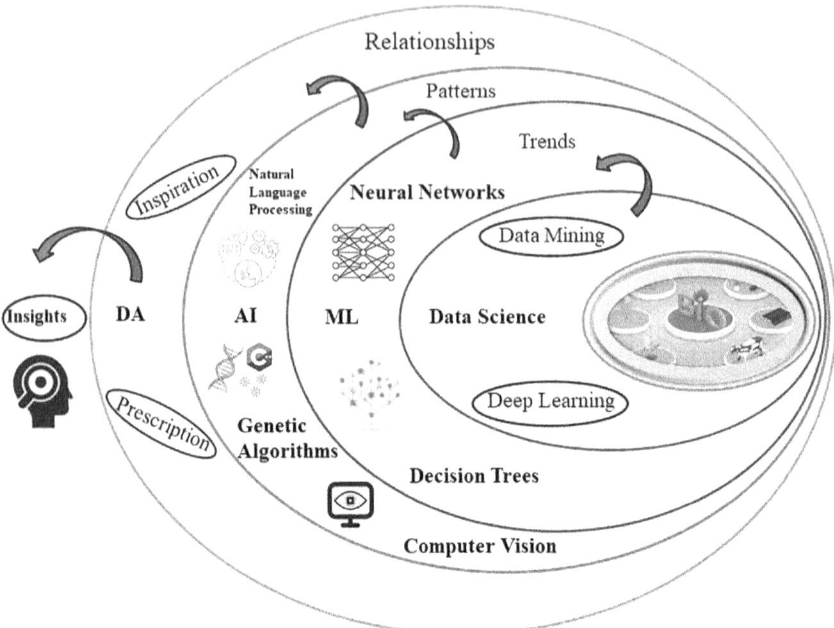

Figure 2.8 Context awareness in IoT systems is made possible by machine learning algorithms, which are essential.

minimizing downtime. In smart grid systems, machine learning algorithms can analyze energy consumption patterns and adjust energy distribution based on real-time demand and supply. Furthermore, machine learning can be applied to optimize resource allocation, scheduling, and control in various CPS domains. By leveraging machine learning techniques, context-enabled CPS can enhance efficiency, reliability, and adaptability, enabling intelligent and autonomous operation in complex and dynamic environments.

2.8 ENERGY-EFFICIENT COMMUNICATION PROTOCOLS FOR IoT-ENABLED CYBER-PHYSICAL SYSTEMS

Energy-efficient communication protocols are crucial for IoT-enabled CPS to optimize energy consumption and extend the lifespan of IoT devices. Energy-efficient communication protocols seek to reduce the energy used during data transmission and reception since IoT devices often have limited resources and battery life. These protocols employ various techniques such as duty cycling, where devices alternate between active and sleep states, reducing the overall power consumption. Additionally, protocols like low-power wide-area networks (LPWANs) utilize long-range communication with low

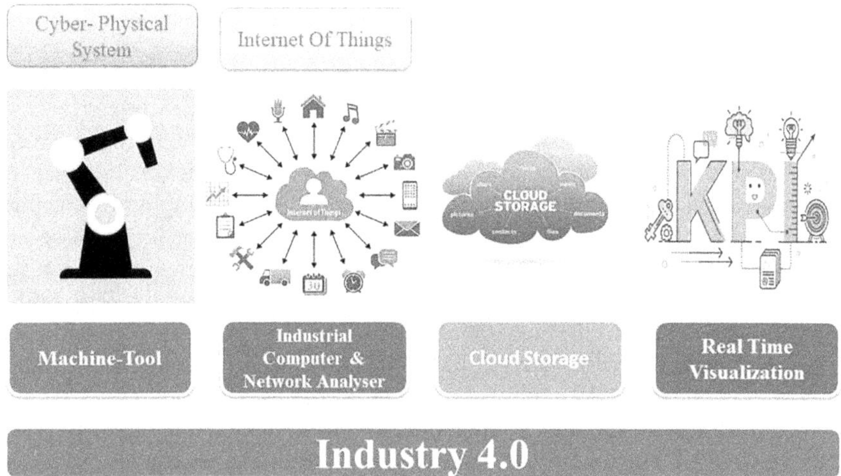

Figure 2.9 Energy-efficient communication protocols are crucial for IoT-enabled CPS.

data rates, enabling devices to operate with minimal power requirements. Other energy-saving techniques include data aggregation, where multiple data packets are combined into a single packet to reduce overhead, and adaptive transmission power control, which adjusts the signal strength based on the proximity of the receiver.

By implementing energy-efficient communication protocols, IoT-enabled CPS can enhance the overall energy efficiency of the system, prolong battery life, and enable the deployment of energy-sensitive applications in various domains such as smart homes, industrial automation, and environmental monitoring.

Energy-efficient communication protocols play a critical role in IoT-enabled physical systems, ensuring optimal energy utilization and prolonging the battery life of IoT devices. These protocols are designed to minimize the energy consumption during data transmission and reception, considering the resource limitations of physical devices. By employing techniques such as duty cycling, where devices alternate between active and sleep states, energy-efficient protocols can significantly reduce power consumption. Moreover, protocols like Zigbee and Bluetooth Low Energy (BLE) utilize short-range communication with low power requirements, enabling devices to operate for extended periods without frequent battery replacements. Additionally, data compression and aggregation techniques are employed to minimize the amount of data transmitted, reducing energy overhead. By implementing energy-efficient communication protocols, IoT physical systems can effectively manage energy resources, improve overall system performance, and support applications that require long battery life, such as wearable devices, environmental monitoring systems, and remote sensing applications.

2.9 SOCIETAL AND ENVIRONMENTAL IMPACT OF IoT-ENABLED CYBER-PHYSICAL SYSTEMS

IoT-enabled CPS has a significant societal and environmental impact, transforming various aspects of our lives. On the societal front, IoT-enabled CPS has improved efficiency and convenience in sectors such as healthcare, transportation, and public safety. Remote patient monitoring and wearable health devices allow for proactive healthcare management, reducing hospital visits and improving patient outcomes. Smart transportation systems enable optimized traffic flow, reducing congestion and improving road safety. Additionally, IoT-enabled CPS has enhanced public safety through surveillance systems, emergency response mechanisms, and disaster management applications. Furthermore, IoT-enabled CPS has a positive environmental impact by promoting sustainability and resource conservation. Smart energy grids leverage IoT sensors and advanced analytics to optimize energy distribution, reduce wastage, and enable renewable energy integration. Energy management systems in smart buildings optimize energy use based on occupancy and environmental factors, lowering their carbon footprints. IoT sensors are used by waste management systems to monitor and improve rubbish collection routes, lowering fuel use and environmental pollution.

The possible detrimental social and environmental effects of IoT-enabled CPS, however, must be taken into consideration. The enormous volume of data that IoT devices gather and send raises privacy and security issues. Maintaining confidence and addressing possible hazards requires protecting personal information and guaranteeing data privacy. Additionally, the e-waste generated by IoT devices poses environmental challenges, requiring proper disposal and recycling measures. Overall, the societal and environmental impact of IoT-enabled CPS is vast and multifaceted. By leveraging the transformative capabilities of IoT, we can create more sustainable and efficient systems that improve our lives while addressing the challenges and risks associated with its implementation.

2.9.1 IoT-Enabled Physical Systems

IoT-enabled physical systems have a profound societal and environmental impact, revolutionizing various aspects of our daily lives. In terms of societal impact, IoT-enabled physical systems enhance efficiency, convenience, and safety in different domains. Smart homes equipped with IoT devices automate and optimize energy consumption, lighting, and security, providing comfort and convenience to residents. Wearable health devices and fitness trackers promote proactive healthcare monitoring, enabling individuals to track and manage their well-being effectively. Additionally, IoT-enabled physical systems contribute to public safety through smart city applications, including intelligent transportation systems, surveillance systems, and emergency response mechanisms. From an environmental perspective, IoT-enabled physical systems have a positive impact on sustainability and resource management. Smart energy grids leverage IoT sensors and real-time data analysis to optimize energy distribution, reduce wastage, and integrate renewable energy sources effectively. Smart agriculture systems enable precision farming techniques, conserving water, minimizing pesticide use, and maximizing crop yields. Waste management systems employ IoT sensors to monitor and optimize waste collection routes, reducing fuel consumption and environmental pollution.

However, it is essential to address potential challenges associated with IoT-enabled physical systems. Strong data protection measures are required owing to privacy and security issues raised by the enormous volume of personal data that IoT devices gather. Additionally, the electronic trash produced by IoT devices may be hazardous to the environment, emphasizing the need of proper disposal and recycling procedures. Overall, IoT-enabled physical systems bring about significant societal and environmental benefits by enhancing efficiency, sustainability, and safety. By embracing these technologies responsibly and proactively addressing associated challenges, we can harness their transformative potential for the betterment of society and the environment.

2.10 HUMAN-MACHINE INTERACTION IN IoT-ENABLED CYBER-PHYSICAL SYSTEMS

Human-machine interaction (HMI) plays a crucial role in IoT-enabled CPS, shaping the way humans interact with intelligent machines and devices. HMI in IoT-enabled CPS focuses on designing intuitive interfaces and user experiences that enable seamless communication and control between humans and machines. These interfaces allow users to interact with IoT devices, access and analyze data, and make informed decisions. HMI designs strive to simplify complex processes, provide real-time feedback, and enable customization and personalization to enhance user engagement and satisfaction. Moreover, voice recognition, gesture-based controls, and augmented reality interfaces are being integrated into IoT-enabled CPS, enabling more natural and immersive interactions. Effective HMI design considers factors such as usability, accessibility, and adaptability to diverse user needs. As IoT-enabled CPS continues to proliferate in various domains, ensuring intuitive and user-friendly HMI will be critical in enabling individuals to harness the full potential of these systems while facilitating effective human-machine collaboration.

2.10.1 Human-Machine Interaction (HMI) Physical Systems

HMI plays a vital role in IoT-enabled physical systems, enabling seamless communication and interaction between humans and the physical devices or systems. In IoT-enabled physical systems, HMI designs focus on providing intuitive interfaces that allow users to monitor, control, and interact with physical devices or objects in their environment. These interfaces can take various forms, such as mobile applications, touchscreens, and voice-activated systems, depending on the context and user requirements. The goal of HMI in IoT-enabled physical systems is to simplify complex tasks, enhance UX, and facilitate efficient and effective control and monitoring. With well-designed HMIs, users can easily access real-time data, adjust settings, and receive feedback from physical devices. This interaction empowers users to make informed decisions and optimize the operation of physical systems. Effective HMI design in IoT-enabled physical systems is essential for maximizing user engagement, usability, and overall system performance while ensuring that individuals can effectively leverage the benefits of these technologies in their daily lives.

2.11 CLOUD COMPUTING AND IoT-ENABLED CYBER-PHYSICAL SYSTEMS

2.11.1 Cloud Computing and IoT-Enabled Cyber-Physical Systems

IoT-enabled CPS relies heavily on cloud computing because it provides scalable infrastructure, storage, and computational capabilities to manage the enormous volume of data produced by IoT devices. Real-time data processing, analytics, and seamless communication are made possible by the combination of cloud computing with IoT. All devices can transmit sensor data to the cloud for storage and analysis, allowing for centralized data management and efficient processing. Cloud computing provides the necessary resources for handling complex analytics, machine learning algorithms, and AI models, enabling intelligent decision-making and predictive capabilities in IoT-enabled CPS. Furthermore, the cloud offers a platform for remote device management, software updates, and system monitoring, ensuring the seamless operation and security of IoT devices. By leveraging the power of cloud computing, IoT-enabled CPS can achieve enhanced scalability, flexibility, and computational capabilities, paving the way for advanced applications such as smart cities, industrial automation, and personalized services.

2.11.2 Cloud Computing and IoT-Enabled Physical Systems

Cloud computing plays a pivotal role in IoT-enabled physical systems, providing scalable and flexible resources to handle the large-scale data generated by physical devices. By integrating cloud computing with IoT, physical systems can leverage the power of the cloud for data storage, processing, and analysis. IoT devices can transmit sensor data to the cloud, enabling centralized data management, real-time monitoring, and advanced analytics. Cloud computing platforms offer robust computing capabilities and storage infrastructure, allowing physical systems to handle complex computations, machine learning algorithms, and predictive models. Moreover, cloud-based services provide remote device management, software updates, and system monitoring, ensuring efficient operation and security of IoT-enabled physical systems. By harnessing the capabilities of cloud computing, IoT-enabled physical systems can achieve scalability, reliability, and cost-effectiveness, enabling advanced applications such as smart infrastructure, environmental monitoring, and precision agriculture. Cloud computing empowers physical systems with enhanced computational capabilities and data-driven insights, driving innovation and efficiency in various domains.

2.12 CONCLUSION: FUTURE DIRECTIONS FOR RESEARCH AND DEVELOPMENT IN IoT-ENABLED CYBER-PHYSICAL SYSTEMS

In conclusion, the field of IoT-enabled CPS has witnessed significant advancements and transformative impacts on various domains. As we look towards the future, there are several promising directions for research and development in this field. First, there is a need for further exploration of advanced communication protocols and network architectures that can address the scalability, reliability, and security challenges of IoT-enabled CPS. Developing efficient and energy-aware communication protocols will be crucial for maximizing the potential of IoT devices in diverse applications. Second, the integration of artificial intelligence and machine learning techniques in IoT-enabled CPS holds immense potential. Advancements in AI algorithms, such as deep learning and reinforcement learning, can enable intelligent decision-making, predictive analytics, and adaptive control in real time. This can lead to enhanced autonomy, optimized resource allocation, and improved system performance.

Third, the development of robust and privacy-preserving frameworks for data management and analytics in IoT-enabled CPS is crucial. As the volume and sensitivity of data collected by IoT devices continue to increase, ensuring data privacy, security, and compliance will be paramount. Future research efforts should focus on developing scalable and efficient data management techniques that can handle the massive influx of data while preserving privacy and maintaining data integrity. Additionally, there is a growing need for interdisciplinary collaborations and standards to drive the adoption and interoperability of IoT-enabled CPS across different industries and domains. Collaboration among researchers, industry experts, policymakers, and stakeholders will be essential to address technical challenges, societal implications, and ethical considerations associated with IoT-enabled CPS.

Lastly, the human aspect of IoT-enabled CPS should not be overlooked. Further research should focus on improving human-machine interaction design, usability, and user experience in order to empower individuals to effectively interact with and control IoT devices. Ensuring that IoT technologies are user-friendly, intuitive, and accessible will be crucial for their widespread adoption and acceptance. Overall, the future of IoT-enabled CPS holds immense potential for revolutionizing industries, enhancing quality of life, and addressing global challenges. By focusing on these future directions for research and development, we can unlock the full potential of IoT-enabled CPS and pave the way for a more connected, intelligent, and sustainable future.

Chapter 3

Design and Implementation of the Distributed Dosimetric System Based on the Principles of IoT

C. Priya

3.1 INTRODUCTION

One of the leading trends in the development of information management systems (IMSs) now is the wide application of the principles of the Internet of Things (IoT). This approach is successful in various fields of science, education, medicine, and industry. It could create favorable conditions for remote work with radiation data due to the use of modern hardware base and new software solutions. The harmfulness of working conditions and territorial distance from educational institutions would cease to be factors that complicate the process of training new specialists and the development of the radiation industry. It is advisable to connect radiometric and dosimetric devices of individual laboratories to the general network using IoT tools, improving the total result of the system parts and providing prompt and reliable access to experiments online. Despite the significant number of scientific works related to building IMSs on the principles of IoT.

The limited use of certain information and physical resources during the pandemic has increased dramatically. Therefore, establishing remote access to the working space is extremely important to support the work of various spheres of human activity. These activities include monitoring radiation conditions of the environment and performing radiometric and dosimetric studies. The work and training of specialists are also complicated by the relatively limited access to closed laboratories engaged in radioactivity research. That is why it is relevant to disseminate dosimetric experiment data in real time by constructing an extensive spectro-dosimetric system with remote access.

3.2 LITERATURE REVIEW AND PROBLEM STATEMENT

This chapter describes the potential of radiological institutions, which can be significantly increased by distributing experimental data on the principles of the IoT, a promising concept of combining physical devices into a single system in order to obtain useful effects as a result of sharing data [1] coming

DOI: 10.1201/9781003406723-3

from separate, geographically branched system elements [2]. However, the issues of combining radiation equipment and laboratories into a system based on the IoT remained unresolved. Owing to this, we shall be able to produce a unique synergistic effect when the integrated indicator would exceed the sum of the research results of individual scientific institutions. This will surely happen due to an increase in the number of researchers [3] who will be able to access experimental information, excluding duplicate measurements, or vice versa, by accelerating the control confirmation of the data obtained [4]. A particularly significant result could be obtained for remote training of specialists in the radiation industry [5].

The architecture, the principle of operation, and the general structure of the spectro-dosimetric system are described in this chapter. The interaction of subsystem components and variants of their interaction with each other are clearly shown. The sequence of interactions between parts of the system and Android application users is shown. An important step was the use of data processing in the cloud service using MATLAB packages. The disadvantages of these subsystems are the power supply of modules in remote locations where there is no network. Based on our research, this chapter describes the functionality of applications using unified modeling languages (UML) diagrams, such as diagrams of Android application precedents and a dosimetric module, where actors are a client and an engineer who have the ability to configure modules and applications of all subsystems. The UML dosimetric module precedent diagram has a series of disadvantages in the absence of part of the detector and cloud service, because there is a challenge of active actors, on a timer. The UML module states diagram reveals the behavior of the module from the moment of power supply to a regular or emergency shutdown and reveals the requests for detection and sending that the module performs. The main issue was to build a class diagram of the software for the data exchange module and the client Android application. Combining the implementation of the work of the microcontroller and software has made it possible to construct a distributed dosimetric experimental data dosimetry system based on semiconductor detectors. The structure scheme of the dosimetric module of the system considers analysis of the main parameters of the detection systems of charged particles, such as sensitivity, the range of measured energies of ionizing radiation, and the signal/noise ratio [6]. Thus, the calculation of the exposure dose and exposure dose capacity of radiation involves a modified algorithm, which is used in development.

Overcoming the corresponding difficulties can be the system architecture and a hardware and software complex that could implement the joint work of its elements based on semiconductor dosimetry with CdTe detectors of nuclear radiation.

3.3 THE AIM AND OBJECTIVES OF THE STUDY

The purpose of this authentic research is to develop an architecture and structural scheme of a distributed dosimetry system based on semiconductor detectors and to conduct a study into the features of its components. The would-be hardware and software solutions could make it possible to demonstrate a full cycle of work with the data from a dosimetric experiment, including remote access to experimental equipment and the building of fundamentally new schemes for the wireless network of radiation monitoring of the environment. To accomplish the aim, the following tasks have been set:

- to devise a general representation of the architecture based on determining the operational principles of the dosimetric system;
- to design components of the dosimetric system and options for their interaction;
- to construct a sequence of executing the processes and sequence of operations of interaction of elements in the system;
- to build UML diagrams of the dosimetric system;
- to prepare a structure scheme of the dosimetric module.

3.4 THE STUDY MATERIALS AND METHODS

The object of research is a distributed information management system for collecting, processing, and distributing dosimetric experimental data based on the principle of the IoT.

An analytical approach has been applied using the MathWorks (USA) ThingSpeak platform, which is well adapted for IoT projects [7]. The ThingSpeak platform enables one to create private or public channels, use the RESTful and MQTT APIs, set the execution time of commands, and send warnings. The service has rich libraries for working with different programming languages and supports Arduino, Particle Photon and Electron, ESP8266, Raspberry Pi, mobile and web applications, Twitter, Twilio, and much more. An important factor is that MathWorks owns the popular MATLAB mathematical package, which is integrated into the ThingSpeak platform. In the study of the information and management system, the mathematical packages MATLAB are used. The use of UML has made it possible to build and investigate the components of the dosimetric system and their interaction variants, which are designed to measure the dose of ionizing radiation. All other components and indicators of interaction of elements in the system to build a model are calculated and demonstrated as follows.

3.5 RESULTS OF STUDYING THE WORK OF A DISTRIBUTED SPECTRO-DOSIMETRIC SYSTEM WITH REMOTE ACCESS

Figure 3.1 depicts the proposed general architecture of the distributed dosimetric system, which is a series of the same type of research (DL) and educational (TL) cells (subsystems). All research centers have sets of hardware

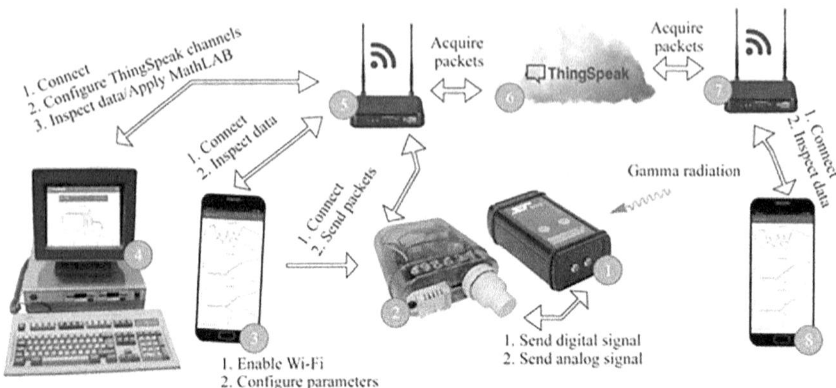

Figure 3.1 Distributed dosimetric system architecture.

and software elements for controlled measurement of radiation parameters of objects or the environment. Data exchange between elements in the subsystem and between subsystems and the analysis of the received information are carried out with the active use of cloud services. An important option of the system is the possibility of remote interference in the measurement process online. As shown in Figure 3.1, it allows an online transmission of initial and processed experimental data to classrooms.

Traditionally, a dosimeter registers the presence of radioactivity, for example, gamma radiation, and transmits digital signals, proportional to the dose rate of ionizing radiation, to the with the Internet via a Wi-Fi router (5); a smartphone (3) can also receive indicators of the parameters of the channel of the cloud service ThingSpeak (6). To select the desired communication channel, there is a configuration page for the plugin second component—module (2) for interpretation, preliminary analysis, and data transmission. The proposed solution is preferable because the module scheme is based on the Wemos D1 R2 (China) board whose core is also a 32-bit microcontroller ESP 8266 (China). The third component is a smartphone, in this case, operated by Android OS (3). It has the developed application installed. With the help of this application, a smartphone user can configure the module via the Wi-Fi interface.

The fourth component (4) is a desktop computer (PC) connected to the Internet via a wired connection to a Wi-Fi router (5). With the help of a web browser, a PC user can go to the page of the selected channel ThingSpeak (6), view the charts, and operate the data using the built-in MATLAB package [9]. This distinction between the functions of the system is considered optimal. Although, if necessary, mobile device users are also capable of such manipulations. Wi-Fi routers (5) and (7) and cloud service ThingSpeak (6) are ready-made solutions and perform standard communication and computing operations. The functions of remote monitoring and process management give us the best measurements of radiation state parameters carried out using the application installed on the remote user's smartphone or tablet (8).

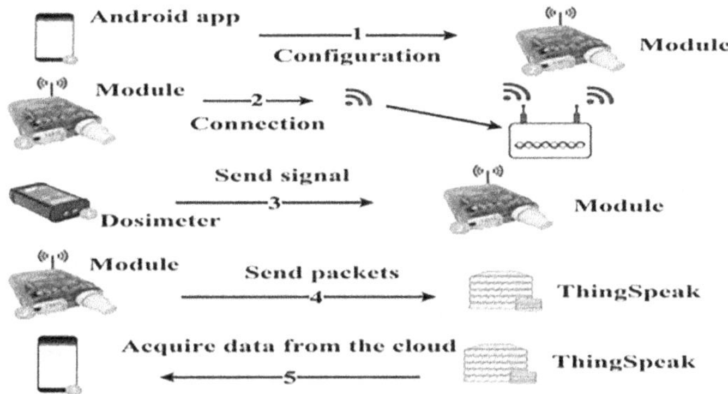

Figure 3.2 The components of the subsystem and their interaction options. The first component of the system (1) is a dosimeter based on the semiconductor CdZnTe detector, microcontrollers STM32F407VGT and ESP 8266 with a Wi-Fi interface. Its structure is similar to that described in [8].

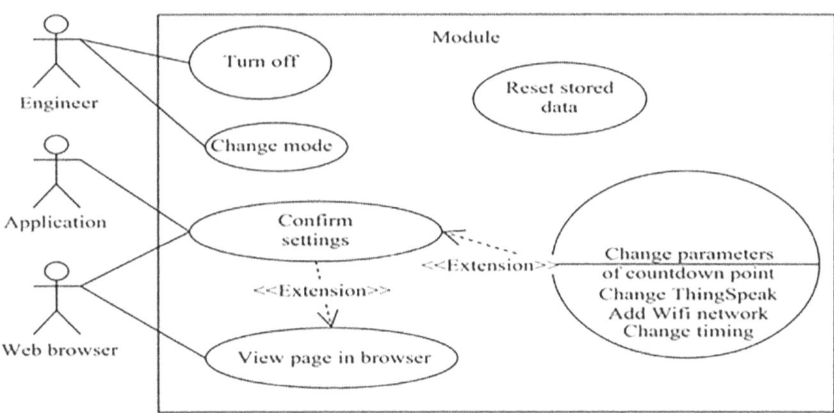

Figure 3.3 Sequence of interaction operations.

Figure 3.3 shows the sequence of interaction operations of parts of the system, which includes the process of deploying and starting the system in a new environment. The process begins with the configuration of the dosimeter and the module of interpretation, analysis, and data transmission using the Android application—step 1. After initialization in the second step, the module connects to the Wi-Fi router to exchange data with the ThingSpeak cloud service. Next—step 3, a dosimeter transmits data on the radiation parameters of the module, which, at step 4, broadcasts preprocessed and compressed data to the cloud. The information processed in the cloud service using MATLAB can be viewed by local and remote users.

In addition, through cloud services, data are exchanged between research teams and, if necessary, broadcasting the measurement process to experts, students, and other interested parties.

3.6 THE UML DIAGRAMS OF ANDROID APPLICATION PRECEDENTS AND DOSIMETRIC MODULE

To determine the functionality of the application by means of UML, a diagram of precedents was built, which is shown in Figure 3.4. Depicted are the actors working with the system, the services it offers, and the types of relations between the system's precedents and its users.

The main actors in the system are a client and an engineer. The secondary actors are parts of the IoT system to which the system connects either directly (module) or with the help of intermediate devices (ThingSpeak). These actors do not initiate the functions of the application.

Actors who initiate the functions of the application are a client and an engineer who inherits all the rights of the client, including the right to configure the module. The client, in turn, can configure the parameters of the application (channel number, number of points depicted, number of data fields) and the absence of an SSID value with the specified access point Wi-Fi password. In addition to the module settings, the settings of the application [10] are also checked. If there are errors in the configuration of the application or module, the user of the application must be notified that the input cannot be confirmed and of the type and place of the error.

The diagram of precedents in the dosimetric module for detecting and transmitting data, shown in Figure 3.5, demonstrates options for interaction of three actors with the functions of the module.

UML diagram of dosimetric module precedents such as Wi-Fi access point authentication parameters (SSID, password), time detection parameters (exposure time, time between measurements, and time between data packets) are set. In addition, the parameters of the ThingSpeak channel used (channel number, API, write and read keys). Changing the mode of operation of the module after accepting the data packet, and deleting previous entries from the module memory, is not mandatory. Changing each of these categories is optional.

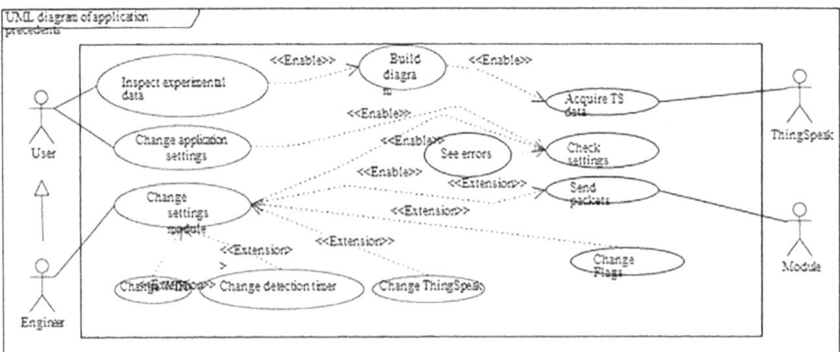

Figure 3.4 Diagram of precedents.

Since some precedents can be called in by hardware using control buttons, in addition to actors–programs, a human actor (engineer) is shown in the diagram.

The web browser configures the module and has the ability to view the web page of the configuration, which the module sends in response to the Protocol (HTTPS) Get request containing the URL specified by the module. Since sending this request is possible not only by clicking the corresponding button on the web page but also manually, using the address bar of the web browser, the precedent of sending settings has a direct association with this actor.

The dosimetric module precedent diagram lacks part of the detector and cloud service. This decision was made due to the lack of a generally accepted designation of precedents, carried out without calling active actors, on a timer. In UML, possible solutions to this problem are the inclusion of an active or passive actor of the "system" type, or a complete absence of associations with such a precedent. However, the introduction of additional actors is not correct due to the lack of intent in the actions of the timer and its belonging to the system. As for the absence of associations with the main actors, this approach is not possible, since for the precedent of "working with data," cloud service and detector act as actors, which can cause misunderstandings when examining the architecture of the system.

3.7 THE UML DIAGRAMS OF MODULE STATUS

Since the module software is an ultimate automaton, an image of its operation is appropriate using the state diagram shown in Figure 3.6.

The module state diagram shows the behavior of the module from the moment of power supply to a regular or emergency shutdown. Its behavior is expressed in the change of system states in response to external and

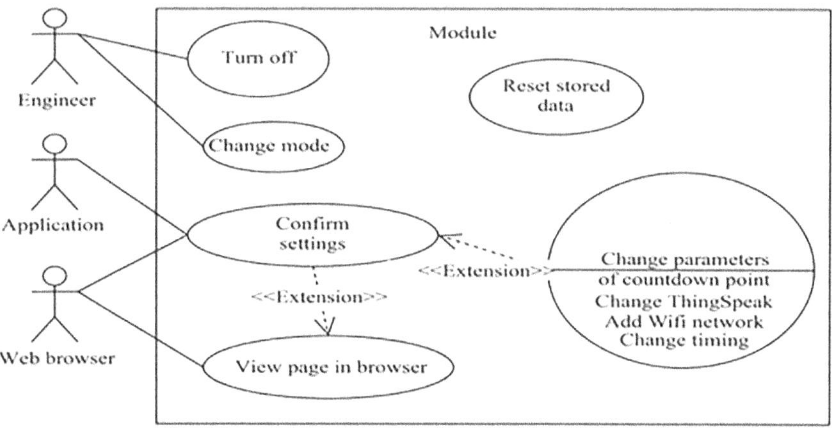

Figure 3.5 UML diagram of dosimetric module precedents.

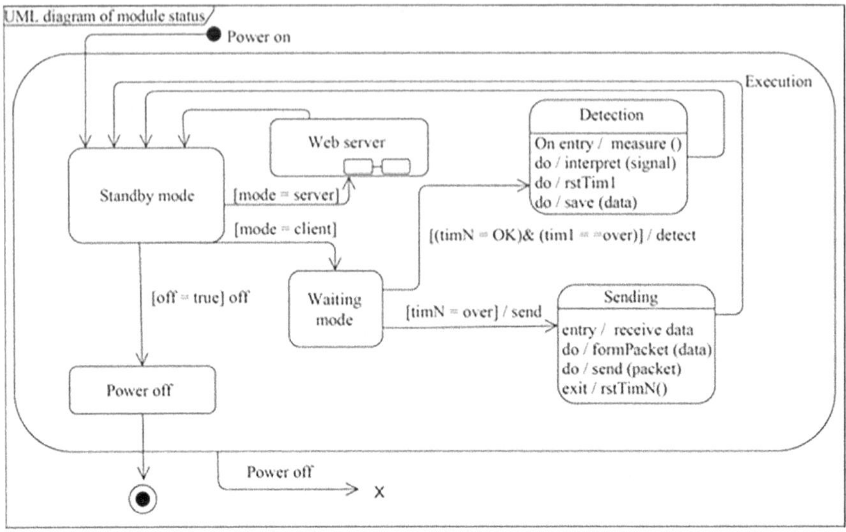

Figure 3.6 UML diagram of module status.

internal events. The general state of the system between the beginning and stopping of its operation is indicated by the rectangle labeled "Running . . ." Because the dosimetric system layout does not work with critical information, maintaining data integrity is not a priority. However, to eliminate the possibility of memory damage, it is necessary to devise a regular method of disabling the module. Unlike the terminate state, the transition to which can occur at any time in the system, the shutdown state involves the completion of the remaining processes. The logic of module operation is built around the cyclic verification of the system mode variable, which can change both when the button is pressed and as a result of selecting the appropriate option through configuration programs. The status of "client" is the main working state of the module. Its work has been demonstrated by waiting, detecting, and sending. After passing the timeout of one detection and the availability of time before sending the packet, the module enters a state of detection. At the same time, it receives, interprets, and writes data to memory. After that, the counter of one detection is reset, and a new cycle of measuring and sending data takes place. After the time before sending, the previously saved data are entered into the packet, the packet is sent to the cloud, the sending counter is reset, and the automaton re-enters the mode check state.

A state diagram includes a web server state that contains a specific number of logical operations, making the program composite. A web server is the part that is responsible for handling signals coming from connected devices. After the first change in the mode of operation to "server," the loop starts the web server, the logic of which is triggered by external interrupts. The interrupt occurs when a new encrypted HTTPS request arrives. For a

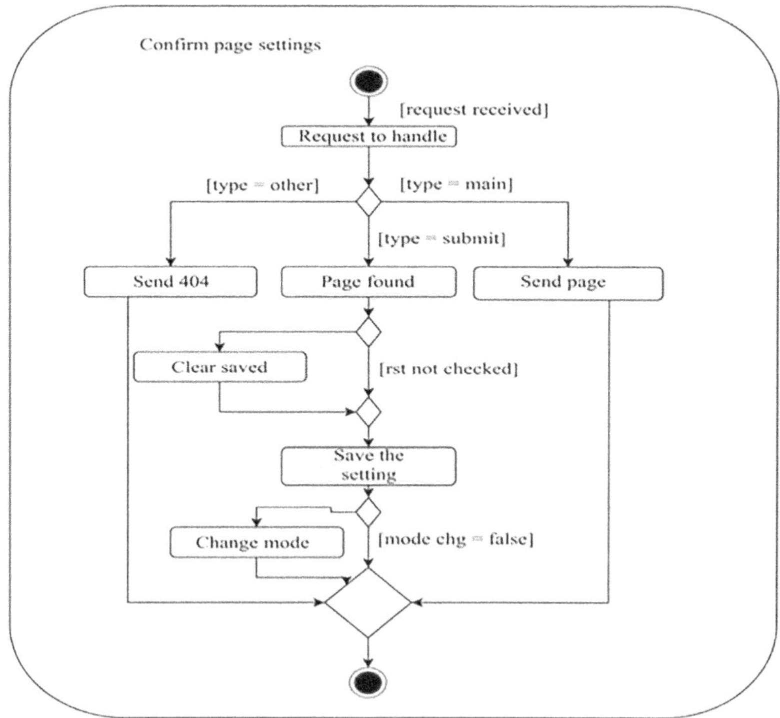

Figure 3.7 UML diagram of HTTPS request handler (module).

more detailed image of the interrupt handler, see Figure 3.7 for a diagram of the activity of the corresponding state of the system.

The activity diagram shows the work of the interrupt handler in the form of a series of events occurring in the event of a request. The entry point is to call the interrupt handler. Significant is the presence of three main query options that correspond to the necessary pages of the elementary website: the main page with configuration forms, a page of successful sending of a message, and a page by default with a message that the entered page is missing on the web server. The central sequence of actions demonstrates the simplified logic of the module configuration process, controlled by the data transmitted in the request. It is worth noting that the removal of previous settings in the activity "Clear saved" should take place before saving new settings, in order to save the current record in the first available cell of the area of energy-independent storage of data [11]. In addition, in this process, it is possible to change the mode of operation of the module when the established mode change attribute is detected. After performing the signal processing, the program enters the main cycle of the program, shown in Figure 3.6. The next stage of work is the formalization of the process described in Figure 3.3 as a diagram of the sequence of interaction of system components.

The diagram of the sequence, shown in Figure 3.8, depicts the joint work of four components of the dosimetric system and two actors: a client of the system (researcher, student) and an engineer with access to laboratory equipment.

3.8 THE DIAGRAM OF SOFTWARE CLASSES OF DATA EXCHANGE MODULE AND CLIENT ANDROID APPLICATION

The description of the dynamics of the dosimetric system layout is enough for the development and program implementation of its algorithms. To complete the development of the system architecture, it is necessary to create diagrams of software classes of the data exchange module and the client Android application that would determine the structural elements, their attributes, and behavior. Figure 3.9 shows the UML diagram of the software classes of the data exchange module. Data exchange module software does not require the creation of a large number of classes. In this process, the client him- or herself can act as an engineer. Separating actors' powers requires authentication capabilities, which involves having a specific infrastructure—a server with authentication data, a unique layout access key, and so on [11].

In addition to the actors, the sequence diagram differs from the described structure of the system by the absence of intermediate devices that make it possible to establish connections with the Internet. Of course, establishing a connection with ThingSpeak without using Wi-Fi access points is not possible for the rest of the system devices—the diagram somewhat simplifies the physical representation of the system structure.

It is worth paying attention to the types of messages exchanged between parts of the system and the focus zone when executing certain commands. One can see that some messages are marked with a filled arrow and will keep the process active until a response from the addressees arrives. Other messages that are asynchronous make it possible to use the system component as soon as one sends the command. In addition, the diagram shows two areas of the cycle. They are responsible for the cyclic detections counting the time between individual detection attempts and the time between sending the data packet to ThingSpeak.

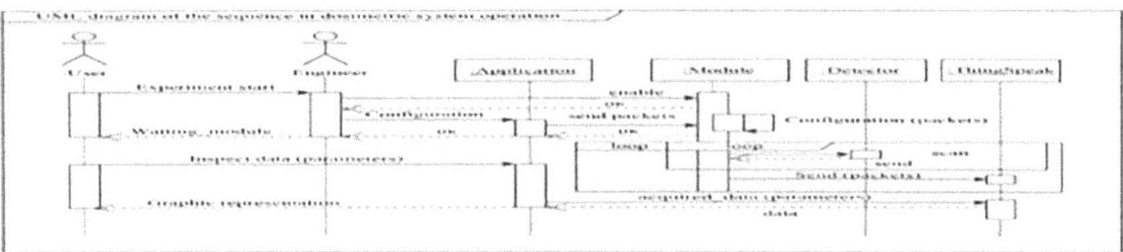

Figure 3.8 Diagram of dosimetric system sequence.

3.9 THE DIAGRAM OF SOFTWARE CLASSES OF DATA EXCHANGE MODULE AND CLIENT ANDROID APPLICATION

The processes that take place in this part of the system are generally implemented without the use of object-oriented programming. Figure 3.9 shows that the classes that the work needs to implement are working classes. The EEPROMWorker class encapsulates all the work with the energy-independent microcontroller memory. It must write SSID pairs and passwords transmitted in HTTPS requests to EEPROM. In addition, it is necessary to store the number of these pairs and the position of the last memory cell. In addition, one needs to be able to clear the memory with a return to the original format and load the stored values into the class responsible for connecting to known Wi-Fi access points. This class, like the rest of the classes absent in this diagram, is part of the finished library. Because of this, their representations in the class diagram are not necessary because they are a "black box" for the system.

More relevant is to construct a class diagram for the client Android application, which is shown in Figure 3.10.

The application class module software diagram shows only classes that need to be developed. To build a graphical user interface, one needs to use a significant number of finished components from different Android libraries. Some links are set at the library level and secret implementation parts. The main class of the application is the MainActivity class. With its initialization, the program begins. The analysis of the implementation of systems,

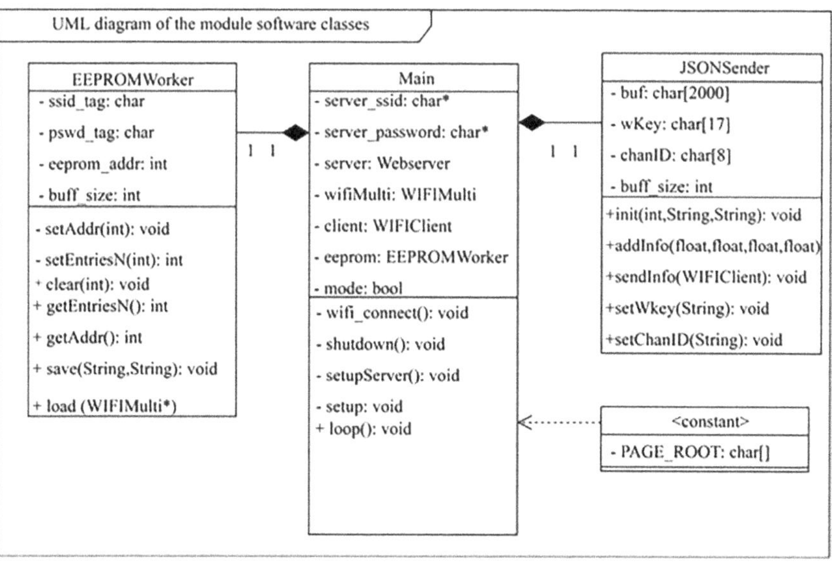

Figure 3.9 UML diagram of the module software classes.

including the elements necessary for this project, make it possible to distinguish a certain number of service classes to ensure the proper operation of the application. To provide the context of the program, a single access point is created in the form of the AppContextProvider class. It makes it possible to transfer the context of the program to other activities besides the main one. The DataViewerSVM class provides the ability to asynchronously access multiple fragments to common variables, namely a data demonstration fragment and an application configuration snippet.

The general data presented here are the data that will act as parameters of the request to the cloud service. Creating such a class is the only option for exchanging data between fragments of the Android application. In addition to these fragments, the system includes a module configuration fragment. Structurally, this fragment is similar to the configuration fragment of the client, but its peculiarity is the ability to form configuration packets and transfer them to the module. All three fragments follow the Fragment class. The snippet change is controlled using the Fragment State Control Adapter (MainFSAdapter).

3.10 BUILDING THE STRUCTURE SCHEME OF THE DOSIMETRIC MODULE OF THE SYSTEM

One of the defining elements of the designed system is a sensor module for measuring the dose of ionizing radiation. Basic parameters of charged particle detection systems, such as sensitivity, range of measured energies of ionizing radiation, signal/noise ratio, and so on, are analyzed. The dosimetric module scheme is built on the basis of a semiconductor CdZnTe detector. Modern microcontrollers STM-32F407VGT and ESP 32 are used to control the module, preprocess data, and organize the wireless Wi-Fi interface according to the protocol IEEE 802.11AH [10].

The structural diagram of the dosimetric module is shown in Figure 3.11.

RS ionizing radiation is recorded by a semiconductor CdZnTe detector D. A signal from the detector, which is proportional to particle energy, is sent to the analog signal processing unit (ASPU), which consists of a preliminary amplifier (RA), fast (RA), and spectrometric (SMA) amplifiers. Pulses from the rapid amplifier are used to count the events of acts of radiation registration. The exposure dose is a complex function of the number of these pulses. The spectrometric channel is used to determine the spectral components of radiation energy and for hardware and software correction of exposure dose measurement results.

Data processing is carried out in a digital processing unit (DSPU). The DSPU hardware includes spectrometric analog-digital converter (SADC), a pulse counter, and a microcontroller (MC).

Calculation of the exposure dose D_e and exposure dose power X_e of gamma radiation is performed according to the modified algorithm used in

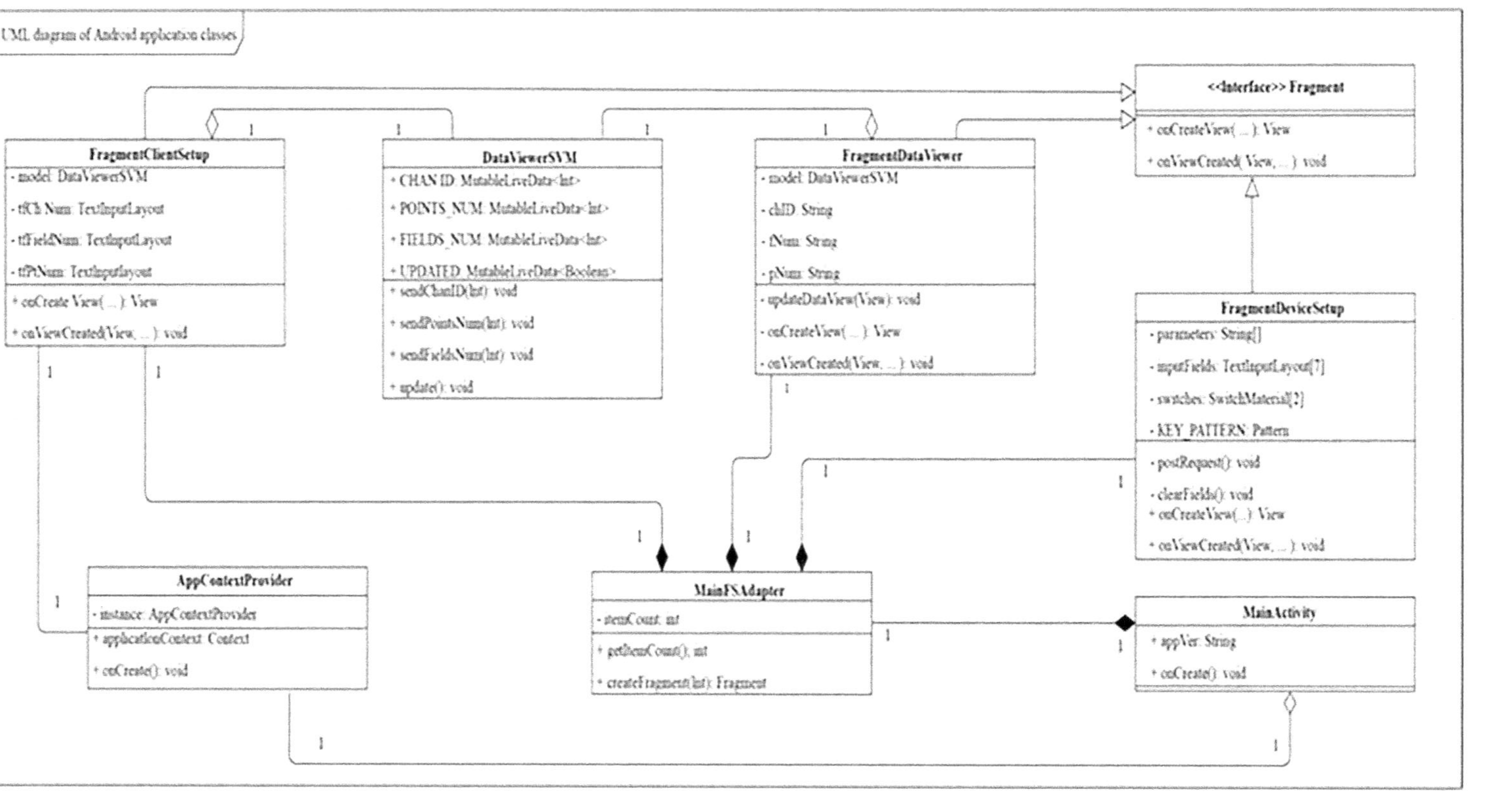

Figure 3.10 UML diagram of Android application classes.

somebody's report. The calculation of these parameters is carried out taking into consideration the peculiarities of the dependence of the detector sensitivity on the energy of the recorded quantum. For a CdZnTe detector, the method of correcting this dependence is proposed in paper [10].

3.11 DISCUSSION OF RESULTS OF STUDYING THE IoT-BASED DISTRIBUTED DOSIMETRIC SYSTEM

Our study of the model of a distributed dosimetric system based on the IoT principles has made it possible to build.

To this end, after recording the spectrum and the number of N_{total} pulses during exposure, the microcontroller program calculates the real exposure.

3.12 DISCUSSION OF RESULTS OF STUDYING THE IoT-BASED DISTRIBUTED DOSIMETRIC SYSTEM

Our study of the model of a distributed dosimetric system based on the IoT principles has made it possible to build. Our hardware and software solutions make it possible to demonstrate the full cycle of the dosimetric experiment. The essence of the functioning of the dosimetric subsystem sequence is in the construction of a certain database of rules for the interaction of the user and intermediate devices of connection with the Internet. The construction of module state diagrams has made it possible to clearly demonstrate the behavior of the module from the moment of power supply and a regular or emergency power outage, but the task is to preserve the integrity of the data and build a regular shutdown of the module. The modular principle of

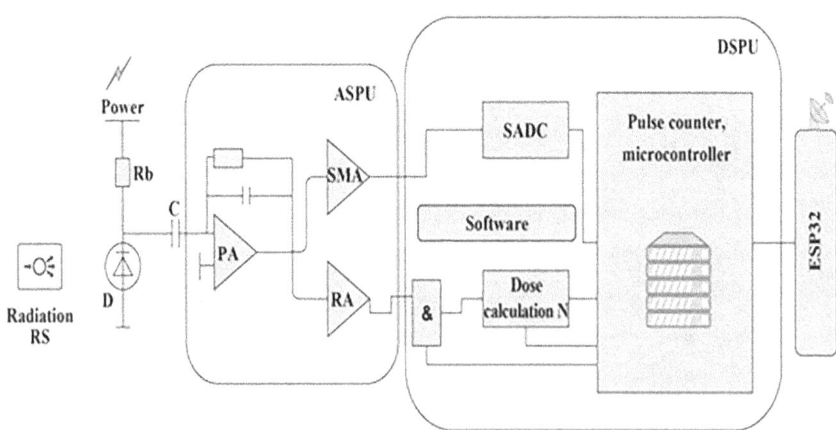

Figure 3.11 Structural diagram of dosimetric module.

organization of the dosimetric system can significantly reduce the number of devices necessary to perform the entire range of tasks due to this approach. The use of cloud services has made it possible to use downloaded data for analysis and visualization. A feature is the architecture of the dosimetric system, which includes two different cells, such as experimental and educational. When solving the main functions, the dynamics of the system, which was presented in the UML diagrams, the change in the mode of operation of the module after the adoption of the data packet, and the deletion of previous entries from the module memory have been described. The disadvantages of the system are the dependence on the presence of a poor signal from the wireless Internet and the inability to request a web server from some countries. The modularity of such a system makes it possible to place mobile points in places where it is needed. Compared to the existing principles of dosimetric systems, it was not possible to adjust the settings online. Developing the Android application allowed us to measure the parameters of the radiation state of the environment from around the world in the presence of the wireless Internet and a smartphone or tablet. The processes taking place in this part of the system are generally implemented without the use of object-oriented programming, which made it possible to develop software using existing libraries and applications. This study could be advanced through the implementation of the developed system in radiation and nuclear-physical processes around the world, using it in the training of new specialists in this field.

3.13 CONCLUSIONS

The general representation of architecture has been proposed on the basis of determining the operational principles of a dosimetric system. The system is a series of the same type of research and training subsystems. The subsystems consist of sets of hardware and software modules for remotely controlled measurement of radiation state of objects or environment.

The components of the system have clearly demonstrated options for their interaction with each other. With the help of a web browser, a PC user can visit the page of the selected ThingSpeak channel, inspect graphs, and operate data using the built-in MATLAB package. Remote monitoring and control functions of the process of measuring radiation state parameters are carried out using the application installed on the remote user's smartphone or tablet.

The smartly devised order of execution of processes and the sequence of operations of interaction of elements in the system make it possible to cover the whole experiment on the principle of the IoT. Thus, the joint work of elements based on semiconductor dosimetry with CdTe detectors of nuclear radiation is organized. Building a sequence of interactions between system parts involves the process of deploying and starting the system in a new environment for the interaction of elements in the system. This interaction begins with the configuration of the dosimeter and the module of interpretation, analysis, and data transmission using the Android application.

The construction of UML diagrams of the dosimetric system makes it possible to fully understand the system and components, as well as methods for acquiring reliable data. Data exchange between elements in the subsystem and between subsystems and analysis of the received information is carried out with the active use of the structural scheme of the dosimetric module. Two-way communication with the cloud with a 15-second loop has been implemented. On a commercial basis, the cycle can be reduced to one second. This is enough for most dosimetric and radiometric studies.

The structural scheme of the dosimetric module, which is designed to measure the dose of ionizing radiation, has been constructed. It has a built-in spectrometric analog-digital converter, microcontroller control, and a communication unit. This makes it possible to accurately measure and calculate the exposure dose and exposure dose rate of gamma radiation due to the new algorithm for correcting the dependence of the detector sensitivity on radiation energy. The communication unit provides wireless connection with laboratory equipment and reliable communication with cloud services.

Measurement by medium charge pulse amplitude is carried out in the energy range from 60 keV to 3 MeV. The advantage of this method in comparison with the traditional method of compensating filters is the absence of a decrease in the sensitivity of measurements in the entire range of the specified energies of the radiation gamma. Due to the hardware–software correction of measurement results, the resolution of the spectrometric channel of 6.5% was reached at the peak of 662 keV full absorption from the reference source Cs-137.

In the future, our architecture, algorithms, and programs could be used for experimental studies of radiation and nuclear-physical processes. In addition, the system elements were useful for remote laboratory work by students during quarantine.

REFERENCES

1. Batura, T. V., Murzin, F. A., Semich, D. F. (2014). Cloud technologies: Basic models, applications, concepts and development tendencies. *Programmnye produkty i sistemy*, 3 (107), 64–72. doi: http://doi.org/10.15827/0236-235x.107.064-072
2. Gubbi, J., Buyya, R., Marusic, S., Palaniswami, M. (2013). Internet of things (IoT): A vision, architectural elements, and future directions. *Future Generation Computer Systems*, 29 (7), 1645–1660. doi: http://doi.org/10.1016/j.future.2013.01.010
3. Ridozub, O., Terokhin, V., Stervoyedov, N., Fomin, S. (2019). Sensor node for wireless radiation monitoring network. *Bulletin of V.N. Karazin Kharkiv National University, Series "Mathematical Modeling. Information Technology. Automated Control Systems"*, 44, 88–93. doi: http://doi.org/10.26565/2304-6201-2019-44-09

4. Finance, G. (2012). *What actor should I use for scheduled use cases?* www.umlchannel.com/en/uml/item/24-use-case-actor-system-timer/24-use-case-actor-system-timer Last accessed: 10.10.2020.
5. Kutnii, V. E., Rybka, A. V., Davydov, L. N. et al. (2021). *Detektory ionizirui-uschikh izluchenii na osnove tellurida kadmiia—tsinka*. Kharkiv: Tipografiia Madrid, 352.
6. Adame, T., Bel, A., Bellalta, B., Barcelo, J., Oliver, M. (2014). IEEE 802.11AH: the WiFi approach for M2M communications. *IEEE Wireless Communications*, 21 (6), 144–152. doi: http://doi.org/10.1109/mwc.2014.7000982
7. Somov, A. S. (2019). *Sbor i vizualizatsiia dannykh s pomoschiu platformy interneta veschei Libelium Waspmote*. Moscow: Skolkovskii institut nauki i tekhnologii, 30.
8. Kumar, S., Tiwari, P., Zymbler, M. (2019). Internet of things is a revolutionary approach for future technology enhancement: A review. *Journal of Big Data*, 6 (1). doi: http://doi.org/10.1186/s40537-019-0268-2
9. MATLAB—MathWorks—MATLAB & Simulink. www.mathworks.com/products/matlab.html
10. Owens, A. (2019). *Semiconductor Radiation Detectors*. CRC Press, 494. Taylor & Francis Group, USA. doi: http://doi.org/10.1201/b22251
11. Zakharchenko, A. A., Nakonechnii, D. V., Shliakhov, I. N., Rybka, A. V., Kutnii, V. E., Khazhmuradov, M. A. (2019). Simulation of energy dependence of CdTe (CdZnTe) gamma-radiation detectors sensitivity. *Tekhnologiia i Innovatsii*, 44, 245–247.

Chapter 4

Next Era of Computing with Cloud Computing and IoT in Different Disciplines

Anjali Naudiyal, Kapil Joshi, Minakshi Memoria, Rajiv Gill, and Ajay Singh

4.1 INTRODUCTION

Internet of Things (IoT) develops the future's architecture generation. It is able to solve new real-time solutions as well as the stream data, with a focus on user-aware, self-aware, and semi-autonomous systems and is also capable of resolving difficulties like filtering, latency, and network-related issues.

The phrase "Next Generation IoT (NGIoT)" first appeared in Europe with the intention of maximizing IoT's usefulness in projects and other areas. Solutions connected to NGIoT aim to lower obstacles to making new discoveries and business models and utilizing expertise. NGIoT has been integrated into a wide range of communication technologies, including blockchain, cloud edge artificial intelligence (AI), and 5G telecommunications networks and services. Humans and a sustainable digital transition are central to many NGIoT projects and funding opportunities. NGIoT collaboration is shown in Figure 4.1.

Horizon Europe is working to provide new opportunities, in addition to completing NGIoT H2020 projects, by starting new IoT-related investigations and new projects both inside and outside of its region. The impact of IoT can determine how effective it is. It has the result that individuals are interacting with other individuals through cyber-physical systems and gaining access to the data produced by connected devices. Using cutting-edge technology and IoT to develop and use IoT, Europe is setting the standard for manufacturing. IoT manufacturing is something that Europe is attempting in many different fields. To manage our society and economy, some of the areas listed later include agriculture, energy, and health care. For a variety of research and innovation groups, industry-driven efforts, and IoT device-related initiatives, NGIoT is exchanging best practices and creating a network of networks. Business first digital global organization is being impacted by NGIoT as an effect of the rapid expansion of the NGIoT world. Networked individuals and automated equipment are travelling down the same route through NGIoT technologies. While some of these developments remain in the development stage, they are beginning to influence company tactics. In the years to come, NGIoT technologies' influence

 DOI: 10.1201/9781003406723-4

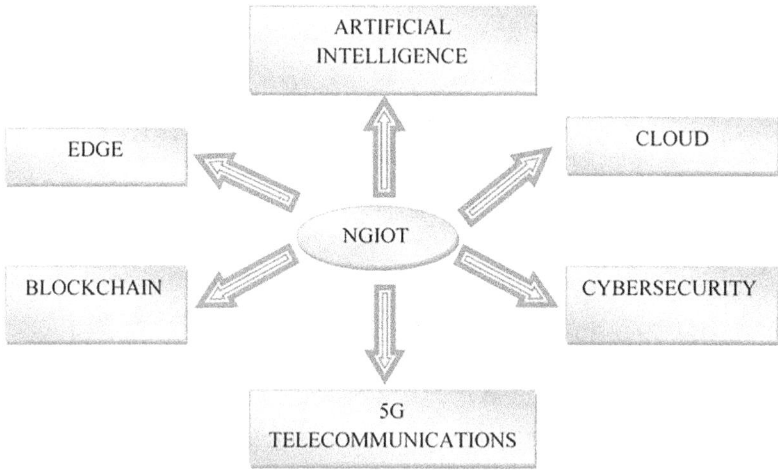

Figure 4.1 NGIoT collaboration.

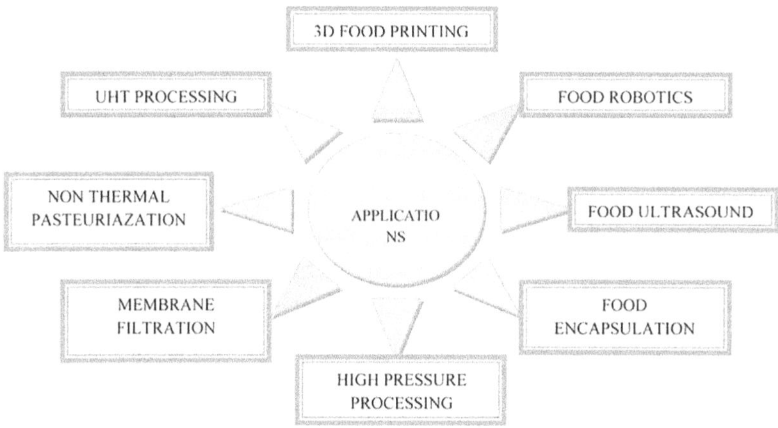

Figure 4.2 Applications of NGIoT.

is anticipated to grow significantly. Various types of applications are introduced in Figure 4.2.

IoT applications are split into five categories as a result of their dependence on networking connections and communication protocols.

1. *Consumer IoT*—This form of IoT is utilized for daily tasks, for example voice assistants for home appliances.
2. *Commercial IoT*—This IoT is utilized in the transportation and healthcare sectors, for example monitoring systems and sophisticated pacemakers.

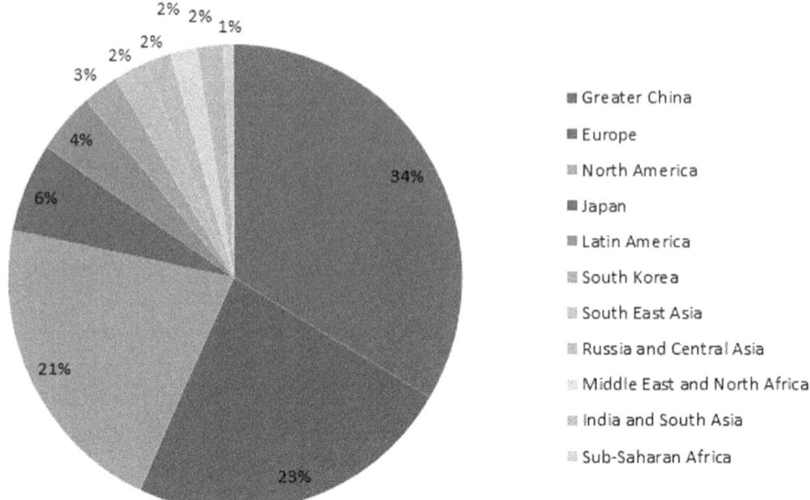

Figure 4.3 Comparison of the regions in which IoT connected devices in Year 2023 and 2030.

3. *Internet of Military Things*—IoMT is utilized in the military, for example human-wearable biometrics for use in battle and surveillance robots.
4. *Industrial Internet of Things*—IIoT is utilized for industrial uses, primarily in the manufacturing and energy sectors, for example smart agriculture, digital control systems, and big data in industry.
5. *Infrastructure Internet of Things*—Mostly employed for networking in intelligent cities, for example sensors and management systems for infrastructure.

4.2 BACKGROUND

The Auto ID Centre used the Internet of Things in 1999 [1] to put the radio frequency identification (RFID) of objects linked through a network together. IoT has facilitated the connection between technology and people. Its goal is to improve the individual's level of comfort, opportunity, and security in life [2, 3]. IoT devices work on the basis that they can be simply plugged into sockets and used.

According to the estimate, there will be 14.3 billion active end points globally in spring 2023, a growth of 18% in the number of IT connections. According to estimates, this rate will rise by 16% in 2023 as IoT-connected devices proliferate at a 16% annual rate, 16 billion active endpoints [4]. Around 27 billion connected devices are predicted to be connected by 2025.

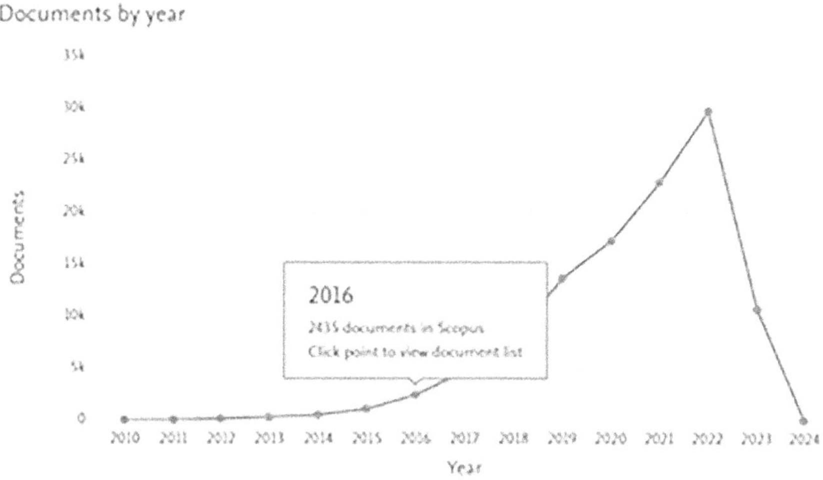

Figure 4.4 Analysis of IoT and cloud computing by year.

IoT devices are arranged in a framework to create a number of important components [5], some of which include the following:

1. Sensor intelligence for monitoring, for example, sound, heat, and light.
2. Uses of intellectualization and algorithms for self-learning of "smart" building subsystems, the creation of a cohesive, secure, sophisticated, and relaxing setting using the items, procedures, and comfortable situations that humans are used to.
3. Standards and protocols like IEEE 1888 for managing sensor network connections and operating them remotely without the involvement of a host (operator). Figure 4.5 shows the analysis of IoT and cloud computing according by author.

The applications of IoT have been made possible by various types of technology: RFID, Bluetooth, internet protocol, wireless sensor networks, artificial networks, Wi-Fi, Zigbee, AI, smart sensors, barcodes, electronic product code, near field communication, and more. Detailed discussion of some technologies are as follows:

1. *Radio-frequency identification*: RFID must be able to detect the environment and objects of the world; a tag with a tiny chip is attached to the object in the real world to interact with the tag using an RFID reader. This can be accomplished by sending the signals that are being generated from the incoming and outgoing signals to the signals database. These signals can be used to identify objects and connect the storage system to the processing centre. Manipal Hospital Bengaluru

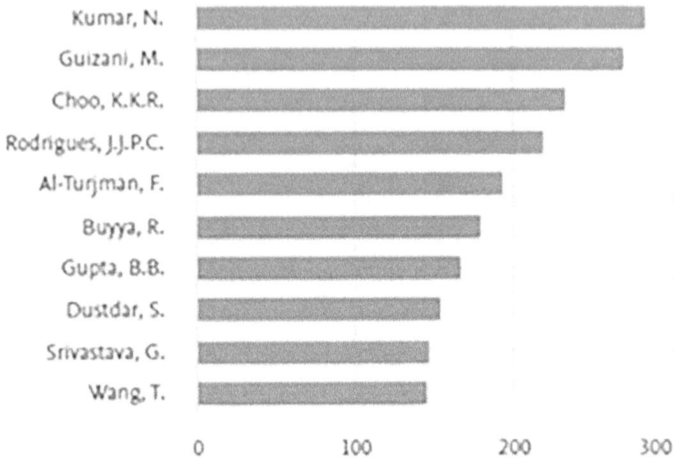

Figure 4.5 Analysis of IoT and cloud computing by author.

employed RFID to control porters, which saved time spent calling nurses who were not essential. Instead of the 33% of Indian nurses using electronic health records, a new system that links hospital information systems or RFID could work to prevent patient misdiagnosis and can be used to reduce medical assessment errors.

2. *Bluetooth*: With the help of this device, the patient can receive the necessary data and continue their progress towards diagnosis during treatment. Bluetooth is a medium for communication connectivity that allows eight to nine devices to be connected through a shared channel called a piconet. Even though there are multiple pieces of medical equipment, they can all still be connected and be very useful in gathering information regarding the patient's issues.
3. *Internet protocol*: The two primary network communication protocols, IPv4 and IPv6, were in use in the 1970s; IPv6 has been dubbed the protocol of the twenty-first century. There are five classes of IPv4 addresses: A, B, C, D, and E. Ipv4 receives 4.3 billion addresses, whereas Ipv6 creates 85,000 addresses.
4. *Zigbee*: Zigbee functions similarly to a network protocol but is mostly utilized to improve wireless sensor networks. The Zigbee protocol IEEE standard is 802.15.4. Data may be made scalable and realizable with the aid of Zigbee, and the data transfer rate can also be decreased if used in biologic medical devices. To communicate and measure massive

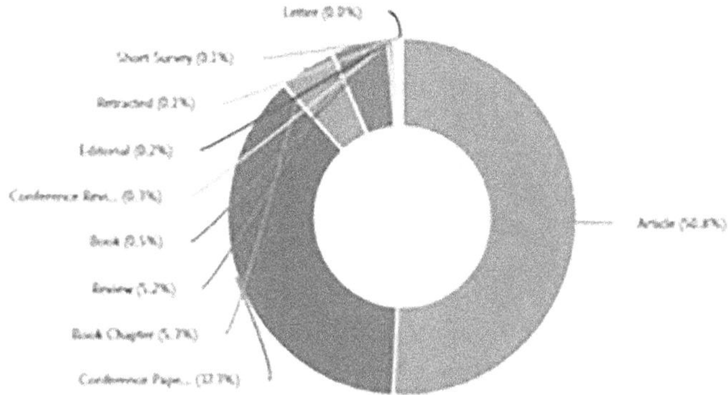

Figure 4.6 Analysis of IoT and cloud computing by type.

volumes of data, Bluetooth and Zigbee are employed. Wireless interfaces like Zigbee and Bluetooth provide a wide range of healthcare applications. The patient or medical professionals can view the data using this wireless interface on the web, a PC, or a mobile device.

5. *Wireless sensor networks*: A WSN allows for supervision of environmental factors as location, speed, pressure, and sound. A key component of the IoT is the WSN, a system of smart gadgets and things. The use of a sensor on a patient or the observation of a drug's impact on it can provide clinicians with information thanks to WSN. WSN is most frequently utilized in the healthcare industry so that doctors can keep an eye on their patients.

4.3 SCOPE OF IoT IN VARIOUS FIELDS

4.3.1 IoT in Communication

The 5G IoT market was estimated to be worth US$ 2.5 billion in 2021, and it is forecast that this amount will increase to US$ 297.1 billion by 2030, or a compound annual growth rate (CAGR) of 70.4% from 2022 to 2030. [6]

To develop 5G business, there is an effort to see opportunities in the fields of manufacturing, energy, healthcare, agriculture, and so on. At present, applications of 5G technology can accelerate connectivity and advance safe driving vehicles and AI enabled robots. Many cases and requirements are being delayed in supply through this technology. The analysis of all the information revealed that the current 5G network operates 10 times more quickly than the LTE network. 5G has proven to be quite advantageous for IoT devices in terms of data sharing and connectivity due to its fast speed.

IoT devices have a real-time network display that allows IoT to control users from remote locations; therefore, their performance is crucial for IoT devices to be commercially successful. 5G IoT can be anticipated to be a crucial development in the global IoT market. IoT applications have enhanced remote heavy machinery control, remote surgery, and data security. The 5G IoT will produce high-quality goods and services for consumers worldwide.

In IoT, numerous protocols are used. Some of them operate at the IP architectural layer, while others do not. In order to fulfil the needs of intra-device and inter-device communication within the network, protocols offer communication services for global networks, local networks, and hybrid networks.

The review data and the communication range are used to categorize the topologies of IoT communication functionality. Signals are said to be mobile when they go from one location to another and reach nodes. Physical and logical mobility, as well as global and local inter- and intra-device mobility, are distinguished. Every type of mobility has a particular communication protocol that is contained in the architecture layer. In both local and global settings, connected mobility is made possible by IP-based protocols for IoT communication. Network ID and host ID are the two halves of the mobility management IP protocol's ID. To transport IoT packets in international connection and to control those using sensors, a border router is required. Managing and regulating sensors involve controlling the data from sensors or routers. The mobility of IoT is solely dependent on the mobility management architecture protocol and does not depend on the ID protocol being used.

Focuses on the fundamental communication between and inside devices, hybrid communication can succeed at both local and wide-ranging communication, it has been regarded as being particularly valuable. In the future, non-IP-based long-distance communication may be successful for device-to-device communication in research.

The criteria were chosen to assess the IoT platform and uphold communication, security, and privacy. Analysis of topological languages for programming, development of applications support, IoT event processing, and support from third parties are all included in this assessment of the IoT platform.

4.3.1.1 Topologies Used in IoT Platform

In our investigation, there were three potential design topologies [7]:

1. Depending on nearby terminal.
2. Communicating directly with a cloud-based server.
3. Combination of a local terminal and a server in the cloud.

4.3.1.2 Hub-Based Topology

It is crucial to have a centralized device in the hub-based architecture so that it can maintain stability in the IoT system. Hub connects IoT devices and cloud-based services, hence it can be compared to a CPU in this regard. After analysing the information that comes in, the centralized hub distributes authority to the equipment that have it. All the equipment used in IoT system are affiliated with the centralized hub [8].

4.3.1.3 Merits

The benefits of having a focal point for centralized communication:

1. As a result of the data being saved in the user's local implementation, users can store their data in any way they see fit and protect it from harmful attackers.
2. While IoT systems can function without an internet connection, newly updated features must be downloaded via the internet. In the hub-based cloud, there is just one point of entry, which both strengthens the protection of the data and makes it easier to access.

4.3.1.4 Demerits

Due to the considerable cost and effort, hub-based topology was abandoned. The user needs a hub or device to set up an IoT system, such as an Apple Home Kit, iPad, and other Apple-related gadgets. Because of how well the hub capabilities perform, this has a significant impact on the solutions. Due to limited internet connection and the hub's vulnerability, which allowed attackers to gain access to the IoT solution's data once, the hub has an impact on security as well [9]. Figure 4.7 depicts an analysis of the IoT and cloud computing implemented devices in a particular subject area.

4.3.1.5 Cloud-Based Topology

IoT solutions are remotely managed by cloud-based servers in a cloud-based topology. In order to allow users to seamlessly integrate their IoT products with cloud services in difficult circumstances, the vendor provides cloud solutions in the IoT platform. By integrating data exchanges, processing, and storage available individually in the conceptual framework of IoT applications, cloud services interact with IoT devices.

4.3.1.6 Merits

Execution time and cloud costs are decreased with the aid of cloud-based topology. Cloud offers quickness. Through cloud-based settings, third-party

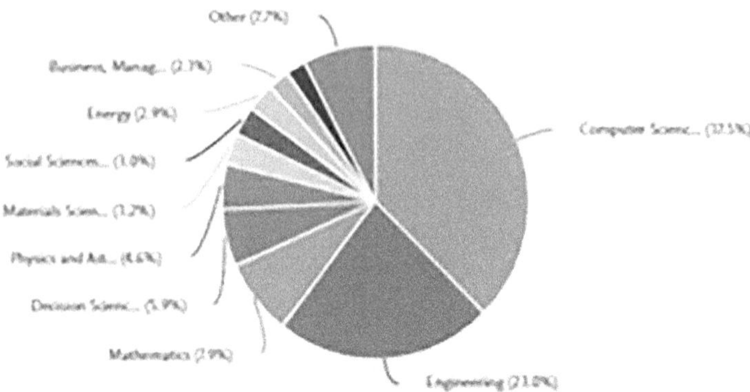

Figure 4.7 Analysis of IoT and cloud computing by subject area.

devices are supported, allowing for greater app customization. Cloud services can be simply implemented by IoT researchers and programmers to boost app customization, unique security, and privacy [10].

4.3.1.7 Demerits

Because IoT solutions cannot operate on an offline internet connection, they need more connectivity and are always necessary. During user communication, the likelihood of data theft increases. In order to protect data transmitted through the IoT network while it is continuously sharing data with the public internet, safety measures have been made available for all devices using an IoT solution. Recent research demonstrates that IoT applications cannot be properly protected by malevolent or artisanal means [11].

4.3.1.8 Hybrid Topology

The most popular hybrid topology is one that uses hybrid architecture, which combines hub-based and cloud-based topologies. IoT solutions with a hybrid topology control, monitor, and share information with cloud services while also managing and controlling devices through a centralized hub. The ability to connect directly to a distant server is present on any device.

4.3.1.9 Merits

IoT solutions cannot lose their performance because alternate approaches are utilized in both disconnected and connected mode. Any time an update is needed, the workload for the storage and processing of data can easily be shared through a hub or cloud.

4.3.1.10 Disadvantages

Information security and privacy are hybrid architecture's key drawbacks. Because each device is able to interact with the cloud-based server separately, security in cloud-based solutions is achievable to some level. As a result, there is less chance that the entire solution will be compromised in the event of an assault on a single device. In hybrid IoT solutions, information must be kept secure on local machines and in the cloud, hence the platform must offer security.

4.3.2 IoT in Health Care

Health care's market size was 180.5 billion dollars in 2021, and according to the Global Internet of Things, it will increase to 960.2 billion dollars by 2030, expanding between 2022 and 2030 at a CAGR of 20.41% [12].

IoT devices in the healthcare industry are posing numerous issues, which include data management, storage, interchange, security, and privacy. In order to speed up the implementation of flexible resources and economies of scale, cloud computing provides services related to computing to the server's database of networking applications and offers data analytics through the internet.

Data analytics are applied to edge services using fog computing, enabling real-time processing while simultaneously securing data privacy and lowering costs [13]. With cloud computing and AI, IoT has the capacity to change how that people receive medical treatment [14]. Figure 4.8 shows the healthcare system related to IoT devices or cloud computing.

IoT makes it less likely for human error to occur when gathering medical data, ensuring that the correct data about the state of the patient is stored. Through the use of the cloud computing paradigm, recently available on-demand computing resources have increased scalability, mobility, and security.

According to a study [15], the capacity to communicate information among people in a precise and organized way has made cloud computing the backbone of the healthcare industry. All of these methods make it simple to identify errors in medical records [16], which has increased the value inside computing clouds and IoT medical fields [17].

A framework with three primary layers is being built for the mobile health monitoring system using cloud computing technology [18].

1. *The several residents control layer and cloud storage*: This platform's foundation, which uses a cloud computing architecture to collect data through sensors, lowers the cost of data and manages data in the several residents control layer and cloud storage.

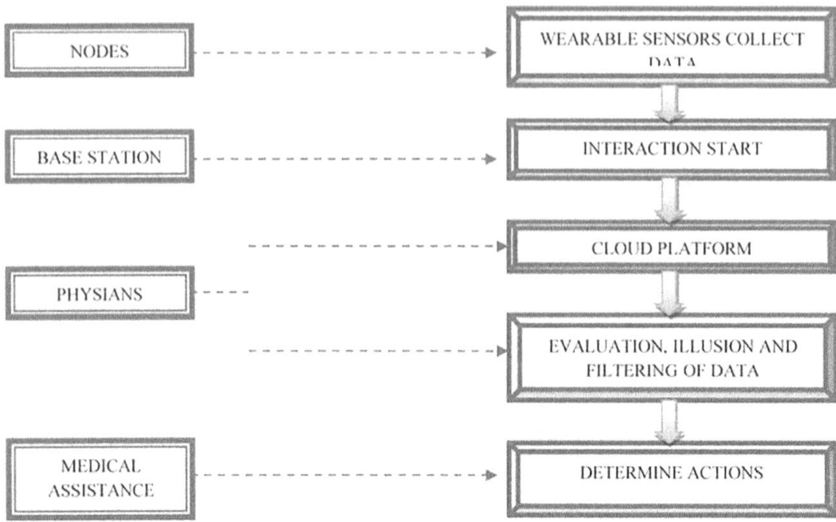

Figure 4.8 Healthcare monitoring system in IoT with cloud.

2. *Medical data augmentation level*: This resolves the issue of several groups of data during data processing in the data annotation layer. The various pieces of hospital equipment are the primary cause of the disparate data information. Various data are produced by various equipment, which makes it difficult to share or comprehend the data. In order to collect fractionated data on the cloud and annotate personal healthcare data, many health professors have gathered to provide live data sets called linked life data (LLD).
3. *The layer for the analysis of medical information*: The previous cloud storage for data for treatment planning for the same disease contributes to the clinical decision and is also analysed, as shown in the image, through the healthcare data analysis layer. This surface utilizes mining techniques to inspire clinical pathways.

IoT applications are broken down into a procedure that consists of three stages via clinical care, remote monitoring, and context awareness, and they play a significant role in many applications in the healthcare sector.

1. *Clinical care*: By developing a system for the care and ongoing observation of patients in hospitals, lives have been saved. The nurse and family members who are caring for the patient receive information about the patient's body from a sensor on the cloud via this monitoring system [19]. In an emergency, patient analysis may be able to save the patient's life.
2. *Remote monitoring*: Modern society is characterized by a fast-paced way of life that prevents people from paying attention to their own or

their family members' health. For children who are already unwell or people who are elderly, daily monitoring is not practical, but a remote monitoring system makes it possible and reduces the need for frequent hospital visits. Remote access sensors aid in diagnosing and enhancing health. This uses people-centric IoT, where a sensor is applied to a specific area of skin to monitor interactions between the heart rate and the effects of medications. In order to monitor patients receiving medical treatments and facilities in remote locations, wireless remote control systems are deployed. In a real-time monitoring system, a patient's heartbeat is monitored if they have a cardiac condition. This is done with the use of a remote medical monitoring unit or monitoring centre, in which the patient was given a sensor from the monitoring centre. Disease is identified by reviewing the data. By using WLAN to send the signals generated by the sensors mounted on the WLAN's body, the nearby medical centre is able to monitor the patient's health and play a significant part in their treatment.

3. *Context awareness*: Context awareness, which plays a significant role in the healthcare industry, is used to identify the patient's health and living circumstances. This allows for the detection of the patient's location as well as the estimation of the patient's surrounding conditions, such as whether it is cold or hot outside or whether the patient's body is experiencing any changes as a result of those conditions. The impact is also discernible. Numerous sensors allow for the detection of the patient's bodily activity, including running, walking, and sleeping abilities. When the patient's condition is known, the appropriate course of treatment can be taken. AI and machine learning are widely used in IoT-connected devices to analyse and increase the value of IoT data, which is then used in healthcare to make the necessary improvements or diagnose problems.

The usefulness of tools composed of AI or machine learning would reach 26.79 billion dollars by 2025, expanding at a CAGR of 29.7% from 2022 to 2025.

4.3.3 IoT in Agriculture

Both the healthcare and agricultural industries have seen significant IoT device growth. This analysis estimates that the market for IoT in agriculture will increase from $27.1 billion in 2021 to $84.5 billion by 2031, with a 12.6% CAGR from 2022 to 2030.

Sensors are used in IoT devices to assess soil moisture, temperature, and plant development parameters.

Figure 4.9 shows an irrigation system in detail. The farmer is able to boost crop productivity, decrease weed growth, and accurately monitor irrigation with the use of data gathered by the sensor. An IoT-based agriculture system has been created so that farmers can easily monitor and protect their crops [20].

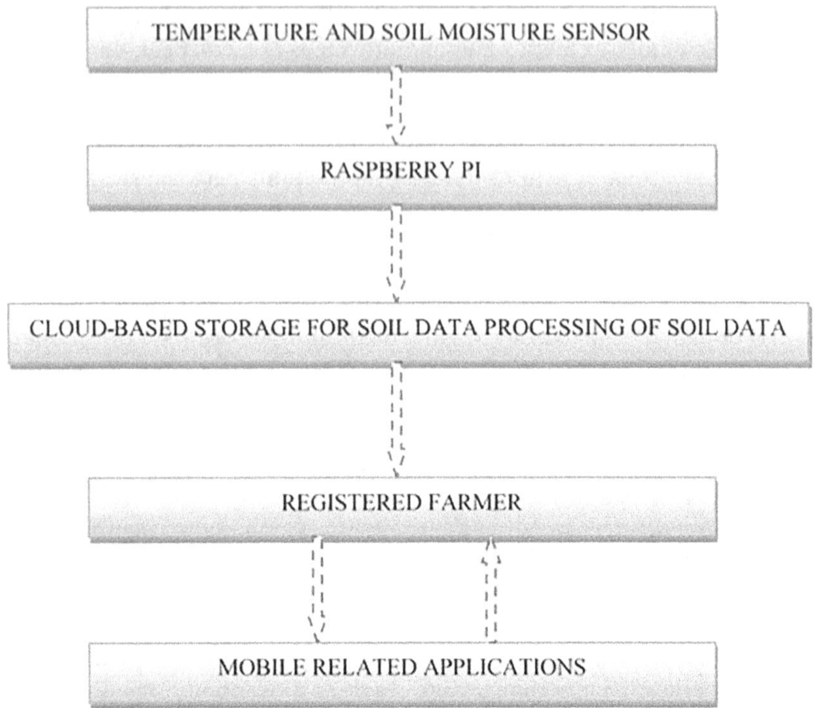

Figure 4.9 Smart irrigation system using mobile application.

4.3.3.1 Raspberry Pi

A multi-switch system that includes essential components such as an Ethernet port, USB connections, an HDMI port, and an SD card slot. The display serial interface (DSI) is crucial as it allows for a permanent connection to the internet and other devices. Its auxiliary capabilities enable connections to peripherals like a mouse, keyboard, camera module, and scanner, facilitating the correct interfacing and transmission of data. With the aid of the operating system, files like apps needed for the project are saved on an 8 GB SD card [21].

1. *DHT11/DHT22 humidity sensor*: The sensor itself monitors the amount of wetness or moisture in the soil with the aid of a Raspberry Pi, and it keeps the information it gathers in the cloud [22].
2. *YL-69 Soil moisture sensor*: Farming and other research facilities utilize the sensor to examine the dielectric properties of voltage fluctuations in water level [23].
3. *Cloud storage*: Soil-related information is stored in a cloud storage repository separate from climate-related data. Utilizing mobile applications, different machine learning techniques are used to gather soil

and climate data. Registered users can examine machine learning predictions and control the level of moisture and humidity for crops.

4.3.4 IoT in Education

IoT in education outlines practical ways to overcome structural and technical obstacles to managing internet services and using information through cloud computing. IoT in cloud computing broadens its application in the real world and has real-world applications of the detail. IoT plays a critical role in streamlining several operations and improving the student experience. On campus, IoT is employed in a variety of applications, including security and energy consumption monitoring.

4.3.4.1 The four Pillars of IoT in Education

To enable the next generation of IoT digital citizens, the four fundamental elements [24] of IoT create a demand for a social effect and appropriate utilization of the knowledge learned in the educational system.

1. *People*: People are also using the internet's functionality to expand and share their experiences in the realm of education. Some people, according to author Alvin Toffler, are able to learn, speak, and learn again. Each individual has a ready node in the network that assists individuals in learning about the work of prominent specialists and contacting those who share their interests and passion. People can talk about the work being done in their region in their linked community in this way. There will be an increase in demand for information sharing from subject-matter experts thanks to streaming and live video. Global education is being expanded through interactive participation in open online courses, including OEDB [25], COURSERA [26], EDX, and massive open online courses (MOOCs) [27]. Connecting online through MOOCs forms, people from every field of life are being developed and connected in an effort to provide them with high-quality education and material.
2. *Process*: In an IoT connected world where process is crucial, data and things collaborate to deliver value. The appropriate processes and methods are used to collect and transmit the appropriate information. By enhancing their need and opportunity for learning and education, learners are driven through the proper method, and resulting in them becoming more proficient. These possibilities will give the students the chance to innovate, which will enable them to succeed in society. Students might increase their comfort and performance by employing proper procedures.
3. *Data*: Items will grow more intelligent as a result of the development of internet-connected items, which will allow us to gather more meaningful information from them. By tagging real-world things, students

can gather information about them, analyse them, and use them for other tasks. With the use of research components, students can monitor data on subjects such as oceanography and climate change, or they can use body-mounted sensors and live webcams to gather information about animals' behaviours in their natural habitats.

4. *Things*: The internet and things were initially connected, but today the internet and people are connected through the usage of sensors on real objects. Students will be able to rapidly track the reading, status, or location of things in real time by connecting IP-enabled sensors in the classroom. It has been discovered through research that studying animals has grown simpler with the use of sensors. Tutors will be enabled to inform students on the most recent results while giving immediate information from their expertise to improve learning.

Cloud computing is a technique whereby virtual machines can access numerous services that are connected to actual hardware in the cloud. Practically expandable storage and retrieval services [28] in public forums are accessible through the cloud.

4.3.4.2 IoT Enabled Education Environment

Those who are struggling academically can connect through cloud computing. IoT is transforming our way of life by enhancing all products. The education environment is extensively covered in the Table 4.2.

Table 4.1 Applications of Cloud Computing in Education

G Suite for Education [29]	A cloud-based total learning management system called Google Classroom via member of Google App for Education product line. Google Classroom is accessible to students on computers, tablets, and smart phones.
Blackboard [30]	Customers, including educational institutions, businesses, and government agencies, are provided with learning, mobility, interaction, and trade software through Blackboard, along with related services.
Knowledge Matters [31]	Knowledge Matters is a significant virtualized online company that uses interactive web-based business simulations to teach critical business ideas to college and high school students.
Coursera [32]	Probably the most well-known educational website. On Coursera, a variety of subjects are available for study.
Microsoft Education Centre [33]	No matter their circumstances, students were expected to be able to continue their education thanks to the Microsoft Education Centre. They enable online learning and give every student the best instruction possible.
Classflow [34]	A cloud-based interactive using a screen-based course delivery system. Without requiring a subscription, Classflow gives clients limitless access to instructional resources.

Table 4.2 A List of the Numerous Educational Environments That Are Used around the World

Smart Education	The goal of intelligent education is to equip students with the knowledge and abilities necessary for success in the marketplace of today. Sensing devices, an Internet of Things architecture, communication links, and user apps are all necessary for the success of smart education [35].
Smart University	A cutting-edge institution includes innovative software and hardware, cutting-edge concepts, current learning strategies, smart teaching methods, and smart classrooms outfitted with the latest technology [36]. Access to a smart university provides a wide variety of international resources, a network-examinable interactive learning environment, and a curriculum that is adaptable to new information. IoT devices for temperature control, security cameras, energy, air conditioning, and building access are commonplace at many institutions.
Smart Classroom	In a "smart classroom," students have access to educational activities using such technology as internet-connected devices, digital screens, and projectors [37].
Smart Teaching	Information transmission using electronic means might be very different from traditional teaching methods. Teachers can stay current on the most recent developments because the material is constantly available and learning is customizable. Access to the real world may be made possible via IoT, but this could make education challenging because it would need to be modified and tailored to accommodate students with different disabilities. Additionally, teaching strategies must be changed to suit children.
Smart Learning	Smart learning is a method of education that uses technology. Smart learning is a process that aids students in their education by focusing on both the subject and the students [38]. The information and communication technology infrastructure is necessary for the intelligence, adaptability, and effectiveness of this technology. Using IoT online education apps is essential for creating an interactive classroom and a globally and locally competitive learning process.
Smart Assessment	By incorporating additional types of evaluation, like focus groups and interviews, smart assessment [39] goes above and beyond the conventional approaches. We must take into account how modern technology has changed how we work in order to provide an appropriate appraisal. Then, when we interact with an ever-expanding IoT ecosystem, the evaluation process changes. The right technologies must be included in modern learning systems in order to record student behaviour for online learning assessments. The student's ability to focus is crucial for evaluating their education, and IoT tools are usable and available in this regard. Exams that are adaptively created and delivered in accordance with each student's desired learning style are technically possible.

4.4 ANALYSIS AND FUTURE SCOPE

Employing cloud IoT will provide the following benefits: safer campus designs, improved operational efficiency, and improved learning experiences and outcomes. The internet is used by an electronic sensor for interaction with the object. IoT [40] refers to systems made of such sensors. It has been estimated that there will be 22 billion IoT-connected devices worldwide by 2025. Every industry has connections, including waste management and electricity distribution, which are proving quite useful in smart homes and allowing for more production to be completed in a shorter amount of time.

The Indian government set a goal of 100 smart cities in 2016, and the state government has invested 200,000 crores to fulfil it by 2024. Among the smart technologies employed in IoT are traffic control, energy harvesting via solar panels on structures and LED streetlights, water conservation via smart metres, and more. IoT-based devices [41] are everywhere, but as IoT [42] data generation grows in volume and complexity, so does the security risk associated with it. For IoT-connected devices, a 5G network is essential because of the high demand. This continues to be difficult for home automation, security, and business alike. IoT has already had a significant impact on a number of industries, from transportation to health care [43], thanks to its capacity to connect a variety of equipment and sensors to the internet.

The transition to edge computing is one of the biggest themes in IoT. Instead of transmitting data to the cloud for analysis, this entails processing it locally on the device. Edge computing can aid with latency reduction, reliability enhancement [44], and privacy and security enhancement. As a result, in the upcoming years, we may anticipate seeing an increase in the number of IoT devices having integrated edge computing capabilities [45].

Smart cities rely on IoT [46]. Cities may improve their infrastructure and citizens' quality of life by integrating a variety of devices and sensors, including waste management systems, environmental monitors, and traffic cameras. Smart waste management solutions, for instance, can lower landfill waste and increase recycling rates while also reducing congestion and improving air quality.

Utilizing blockchain technology to improve security and privacy is a new trend in IoT. Blockchain enables the decentralized, tamper-proof storage of IoT [47] device data, enhancing security and reducing hacker access. By allowing data to be exchanged only with parties who have been given permission, the blockchain can also help to preserve user privacy.

Figure 4.10 depicts the market size of IoT device implementation in particular areas. IoT has the potential to close the gap between the real world and the digital one, with significant and realistic ramifications for the economy and society [48–52]. IoT could be used by the manufacturing sector to boost operational effectiveness and physical asset management. Most importantly, IoT might be a key component for assuring successful digital transformation. The global blockchain IoT market is anticipated to

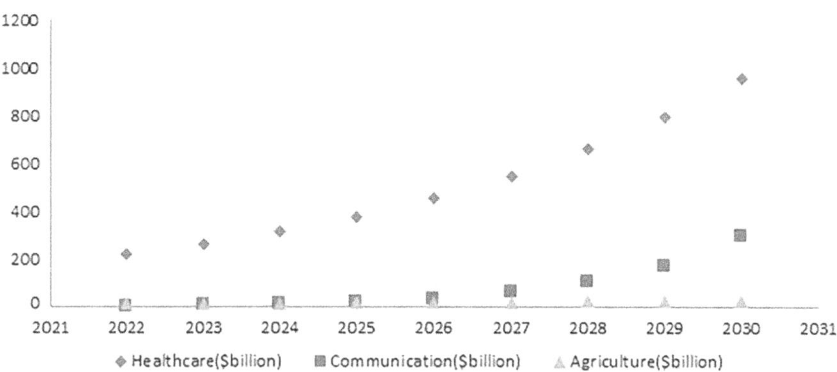

Figure 4.10 A comparison of global market size of IoT device implementation in particular areas.

reach $3.021 billion by 2025, expanding at a CAGR of 93.7% from 2020 to 2025. According to the prediction for IoT-connected devices, by 2030, IoT use cases in standardized production environments could generate between $1.4 trillion and $3.3 trillion. The value creation potential of IoT in manufacturing would prioritize optimization of different operations for greater efficiency [53–56]. As a result, operation-management applications of IoT could make up almost 32% to 39% of economic value generated through IoT. By 2030, IoT applications for optimizing operation management could help in generating between $0.5 trillion and $1.3 trillion in economic value.

4.5 CONCLUSION

The number of IoT devices connected today and into the future around the world is summarized in this chapter. Growing awareness regarding how the Internet of Things is advancing technology is the goal of this chapter. The potential for IoT is essentially limitless, ranging from blockchain and smart cities to cloud computing and AI. In the years to come, we can anticipate even more interesting advancements as businesses and developers keep on innovating and push the limits of what is possible. With the help of smart watches and fitness trackers, the consumer IoT offers the potential to enhance comfort and quality of life. They support the gathering and analysis of information on a user's physical activity and sleep hygiene. The expansion of IoT paves the way for the transition to Industry 4.0 and the development of systems that would not be possible without the greatest possible level of mechanization (intelligence) and direct communication between devices, sensors, and people. The use of smart devices and controllers enhances management effectiveness and the environmental security regime. The result is a decrease in risks and errors and an increase in competition. It improves security, data integrity, and education policies with the aid of successful IoT implementation in education. Cost is also a greater challenge in this field.

REFERENCES

1. K. Ashton, "That internet of things," *RFID Journal*, vol. 22, pp. 97–114, 2009. Accessed: September 26, 2018. [Online]. www.rfidjournal.com/articles/view?4986.
2. K. A. Palaguta, I. S. Shubnikov, and A. L. Safonov, "Handbook of the module smart house," *Kniga*, p. 184, 2016.
3. M. N. Sokolov, K. A. Smolyaninova, and N. A. Yakusheva, "Internet of things security problems: Overview," *Cybersecurity Issues.*, vol. 5, no. Special issue.
4. https://iot-analytics.com/number-connected-iot-devices/#:~:text=In%20 2022%2C%20the%20market%20for,27%20billion%20connected%20 IoT%20devices.
5. G. Quandeng, "Construction and strategies in IoT security system," *Green Computing and Communications (GreenCom) IEEE International Conference on and IEEE Cyber, Physical and Social Computing*, 2013, pp. 1129–1132. doi: 10.1109/GreenComiThings-CPSCom.2013.195.
6. www.precedenceresearch.com/5g-iot-market
7. A. Riahi, Y. Challal, E. Natalizio, Z. Chtourou, and A. Bouabdallah, "A systemic approach for IoT security," *2013 IEEE International Conference on Distributed Computing in Sensor Systems*, May 2013, pp. 351–355. doi: 10.1109/ DCOSS.2013.78.
8. OpenHAB Community, Openhab Documentation, 2017. http://docs.openhab. org/index.html. Accessed: March 9, 2020. [Online].
9. A. K. Sikder, L. Babun, H. Aksu, and A. S. Uluagac, "Aegis: A context-aware security framework for smart home systems," *Proceedings of the 35th Annual Computer Security Applications Conference (ACSAC 2019)*.
10. L. Babun, Z. B. Celik, P. McDaniel, and A. S. Uluagac, "Real-time analysis of privacy-(un)aware IoT applications," *Proceedings on Privacy Enhancing Technologies (Po/PETS)*, vol. 2021, no. 1, 2021.
11. Z. B. Celik, L. Babun, A. K. Sikder, H. Aksu, G. Tan, P. McDaniel, and A. S. Uluagac, "Sensitive information tracking in commodity IoT," *USENIX Security Symposium*, Baltimore, MD, August 2018. www.usenix.org/system/files/ conference/usenixsecurity18/sec18-celik.pdf.
12. www.precedenceresearch.com/internet-of-things-in-healthcare-market.
13. H. L. Truong, and S. Dustdar, "Principles for engineering IoT cloud systems," *IEEE Cloud Computing*, vol. 2, pp. 68–76, 2015. [Google Scholar] [CrossRef].
14. D. L. Minh, A. Sadeghi-Niaraki, H. D. Huy, K. Min, and H. Moon, "Deep learning approach for short-term stock trends prediction based on two-stream gated recurrent unit network," *IEEE Access*, vol. 6, pp. 55392–55404, 2018. [Google Scholar] [CrossRef].
15. N. Sultan, "Making use of cloud computing for healthcare provision: Opportunities and challenges," *International Journal of Information Management*, vol. 34, pp. 177–184, 2014. [Google Scholar] [CrossRef].
16. S. R. Islam, D. Kwak, M. H. Kabir, M. Hossain, and K. S. Kwak, "The internet of things for health care: A comprehensive survey," *IEEE Access*, vol. 3, pp. 678–708, 2015. [Google Scholar] [CrossRef].

17. A. Darwish, A. E. Hassanien, M. Elhoseny, A. K. Sangaiah, and K Muhammad, "The impact of the hybrid platform of internet of things and cloud computing on healthcare systems: Opportunities, challenges, and open problems," *Journal of Ambient Intelligence and Humanized Computing*, pp. 1–16, 2017. [Google Scholar] [CrossRef].
18. B. Xu, L. Xu, H. Cai, L. Jiang, Y. Luo, and Y. Gu, "The design of an m-health monitoring system based on a cloud computing platform," *Enterprise Information Systems*, vol. 11, pp. 17–36, 2017. [Google Scholar] [CrossRef].
19. D. H. Gustafson et al., "Reducing symptom distress in patients with advanced cancer using an e-alert system for caregivers: Pooled analysis of two randomized clinical trials," *Journal of Medical Internet Research*, vol. 19, no. 11, pp. e354, 2017.
20. Nikesh Gondchawar, and R. S. Kawitkar, "IoT based smart agriculture," *International Journal of Advanced Research in Computer and Communication Engineering*, vol. 5, no. 6, June 2016.
21. J. Gutierrez, J. F. Villa-Medina, A. Nieto-Garibay, and M. A. Porta-Gandara, "Automated irrigation system using a wireless sensor network and GPRS module," *IEEE Transactions on Instrumentation and Measurement*, vol. 63, no. 1, pp. 166–176, 2014.
22. K. V. Grace, S. Kharim, and P. Sivasakthi, "Wireless sensorbased control system in agriculture field," *Proceedings of the 2015 Global Conference on Communication Technologies (GCCT)*, Thuckalay, India, 2015.
23. E. Gajendran, S. B. Prabhu, and M. Pradeep, "An analysis of smart irrigation system using wireless sensor," *Multidisciplinary Journal of Scientific Research &Education*, vol. 3, no. 3, pp. 230–234, 2017.
24. www.cisco.com/c/dam/en_us/solutions/industries/docs/education/education_internet.pdf.
25. www.precedenceresearch.com/iot-in-agriculture-market.
26. www.coursera.org/.
27. www.edx.org/.
28. C. Gong, J. Liu, Q. Zhang, H. Chen, and Z. Gong, "The characteristics of cloud computing," *2010 39th International Conference on Parallel Processing Workshops*, IEEE, 2010, pp. 275–279.
29. https://classroom.google.com/.
30. www.blackboard.com/en-apac.
31. https://knowledgematters.com/.
32. www.coursera.org/.
33. https://education.microsoft.com/en-us.
34. https://classflow.com/.
35. D. Mohanty, "Smart learning using IoT," *International Journal of Engineering Research and Technology*, vol. 6, no. 6, pp. 1032–1037, June 2019.
36. V. L. Uskov, J. P. Bakken, R. J. Howlett, and L. C. Jain, "Smart universities: Concepts, systems, and technologies," *International Conference on Smart Education and Smart ELearning*, Springer, Cham, 2018, pp. 1–421.
37. Shrinath S. Pai et al., "IoT application in education," *International Journal of Advance Research, Ideas and Innovations in Technology*, vol. 2, no. 6, pp. 20–24, 2017.
38. D. Gwak, "The meaning and predict of smart learning," *Proceedings of the Smart Learning Korea*, 2010.

39. D. A. Aljohany, R. Mohamed, and M. Saleh, "ASSA: Adaptive E-learning smart students assessment model," *International Journal of Advanced Computer Science and Applications*, vol. 9, no. 7, pp. 128–136, 2018.
40. M. Diwakar, K. Sharma, R. Dhaundiyal, S. Bawane, K. Joshi, and P. Singh, "A review on autonomous remote security and mobile surveillance using internet of things," *Journal of Physics: Conference Series*, Vol. 1854, p. 012024, 2021. doi: 10.1088/1742-6596/1854/1/012034.
41. P. Iyappan et al., "A generic and smart automation system for home using internet of things," *Bulletin of Electrical Engineering and Informatics*, vol. 11, no. 5, 2022.
42. M. Memoria et al., "An internet of things enabled framework to monitor the lifecycle of Cordycepssinensis mushrooms," *International Journal of Electrical and Computer Engineering (IJECE)*, vol. 13, no. 1, pp. 1142–1151, 2023.
43. R. Mishra, K. Joshi, and D. Gangodkar, "Wireless communications network and mobile computing using blockchain in distributed internet of things," *2022 11th International Conference on System Modeling & Advancement in Research Trends (SMART)*, Moradabad, India, 2022, pp. 832–836. doi: 10.1109/SMART55829.2022.10047182.
44. R. Kumar, H. Anandaram, K. Joshi, V. Kumar, J. Reshi, and R. K. Saini, "Internet of things (IoT) applications and challenges: A study," *2022 7th International Conference on Computing, Communication and Security (ICCCS)*, Seoul, Republic of Korea, 2022, pp. 1–6. doi: 10.1109/ICCCS55188.2022.10079508.
45. S. Ghildiyal, K. Joshi, P. Kanti, M. Memoria, A. Singh, and A. Gupta, "A healthcare architecture based on energy-efficient cloud computing," *2022 7th International Conference on Computing, Communication and Security (ICCCS)*, Seoul, Republic of Korea, 2022, pp. 1–6. doi: 10.1109/ICCCS55188.2022.10079256.
46. R. Kumar, S. Kathuria, R. K. Malholtra, A. Kumar, A. Gehlot, and K. Joshi, "Role of cloud computing in goods and services tax (GST) and future application," *2023 International Conference on Sustainable Computing and Data Communication Systems (ICSCDS)*, Erode, India, 2023, pp. 1443–1447. doi: 10.1109/ICSCDS56580.2023.10104597.
47. Pradeepto Pal, Devender Singh, Rajesh Kumar, RajatBalyan, Anita Gehlot, Rajesh Singh, Harishchander Anandaram, Shaik Vaseem Akram, and Kapil Joshi, "Internet of things and cloud server-based indoor talking plant," *SSRG International Journal of Electrical and Electronics Engineering*, vol. 10, no. 4, pp. 57–69, 2023.
48. R. Kumar, M. Memoria, and A. Chandel, "Performance analysis of proposed mutation operator of genetic algorithm under scheduling problem," 2020. doi: 10.1109/ICIEM48762.2020.9160215.
49. A. Rawat, S. Ghildiyal, A. K. Dixit, M. Memoria, R. Kumar, and S. Kumar, "Approaches towards AI-based recommender system," *2022 International Conference on Machine Learning, Big Data, Cloud and Parallel Computing (COM-IT-CON)*, 2022.
50. S. Tyagi et al., "Role of IoT and blockchain in achieving a vision of metropolitan's digital transformation," *2022 International Conference on Machine Learning, Big Data, Cloud and Parallel Computing (COM-IT-CON)*, 2022.
51. A. Rawat, H. Maheshwari, M. Khanduja, R. Kumar, M. Memoria, and S. Kumar, "Sentiment analysis of Covid19 vaccines tweets using NLP and machine learning classifiers," *2022 International Conference on Machine Learning, Big Data, Cloud and Parallel Computing (COM-IT-CON)*, 2022.

52. R. Kumar et al., "Analyzing the performance of crossover operators (OX, OBX, PBX, MPX) to solve combinatorial problems," *2022 International Conference on Machine Learning, Big Data, Cloud and Parallel Computing (COM-IT-CON)*, vol. 1, pp. 817–821, 2022.
53. A. Mohammed et al., "Data security and protection: A mechanism for managing data theft and cybercrime in online platforms of educational institutions," *2022 International Conference on Machine Learning, Big Data, Cloud and Parallel Computing (COM-IT-CON)*, 2022.
54. N. Ansari et al., "A critical insight into the impact of technology in transformation of tourist business into smart tourism," *2022 International Conference on Machine Learning, Big Data, Cloud and Parallel Computing (COM-IT-CON)*, 2022.
55. R. Kumar et al., "A review of memetic algorithm and its application in traveling salesman problem," *International Journal of Emerging Technology*, vol. 11, no. 2, pp. 1110–1115, 2020. [Online]. www.scopus.com/inward/record.uri?eid=2-s2.0-85096815177&partnerID=40&md5=f411b718d77f7b6fa0f05e4317c1948d.
56. R. Kumar et. al., "Analysis of available selection techniques and recommendation for memetic algorithm and its application to TSP," *International Journal of Emerging Technology*, vol. 11, no. 2, pp. 1116–1121, 2020. [Online]. https://www-scopus-com-uttaranchaluniversity.new.knimbus.com/inward/record.uri?eid=2-s2.0-85096815344&partnerID=40&md5=ab8d02f3bad0745cb7abef2a46202a15.

Chapter 5

Design and Development of Secure Mobile Social Network with IoT

Anubha Jain, Priyanka Verma, Usha Badhera, and Pooja Nahar

5.1 INTRODUCTION

A mobile social network (MSN) refers to a social networking platform or service that is specifically designed and optimized for mobile devices, such as smartphones and tablets. It allows users to connect, communicate, and interact with others through social media applications or mobile-optimized websites. MSNs, through mobile devices, offer real-time information services to users on social networks. Mobile social networks have transformed how individuals engage and communicate with each other by enabling users to instantly express their ideas, emotions, and experiences with others.

Increasing social relationships using social platforms like Facebook, Twitter, Instagram, LinkedIn, and others is known as social networking. It brings individuals together for social and professional interactions, allowing them to converse, share ideas and interests, and create new connections. This practice facilitates communication amongst individuals from diverse geographic regions. The ease of use of social networking platforms these days has contributed to their exponential growth in popularity and user numbers.

The fundamental architecture of MSNs will be further discussed later in the chapter. Social networking encompasses general social networks, professional networking, media sharing networks, gaming networks, interest-based networks, microblogging, messaging platforms, and others. *General social networks* are for broad social interaction, cater to a wide range of users, and include Facebook and Instagram. *Professional networks* focus on connecting professionals and facilitating career-related interactions and include platforms like LinkedIn. *Media sharing networks* prioritize the sharing and discovery of multimedia content such as photos, videos, and music. Users can upload, share, and interact with media content through YouTube, Pinterest, and Snapchat. *Microblogging* platforms allow users to share short-form content, typically limited to a certain number of characters or words. Users can post updates, share links, and engage in conversations. Examples include Twitter and Pinterest. *Interest-based networks* focus on connecting users based on shared interests, hobbies, or specific niches. Users can join communities or groups centered on their interests and engage in discussions.

DOI: 10.1201/9781003406723-5

Examples include Reddit and Quora. *Messaging apps* go beyond basic communication functionalities and incorporate social networking features. They allow users to connect, share content, and engage in social interactions alongside messaging. Examples include WhatsApp, Facebook Messenger, and WeChat.

These platforms have become more effective with the integration of IoT, which offers a variety of new features and capabilities to improve user experience and has opened up new social networking opportunities. IoT has a significant impact on MSNs, enabling them to offer enhanced connectivity, personalization, real-time interactions, data analytics, and improved security. These contributions have made MSNs more powerful and effective in connecting people and devices and have opened up new possibilities for social networking in the digital age. However, because of their high levels of independence and accessibility, these MSNs are susceptible to a number of privacy and security issues, including phishing, data breaches, theft of personal information, spreading rumors, and other forms of misuse.

This chapter delves into the realm of MSNs and explores the potential of integrating IoT technologies to enhance its security. The chapter is divided into several sections, each focusing on a key aspect of the topic. It begins with an overview of MSN and its architecture, followed by an examination of the security concerns associated with MSN. Subsequently, the chapter explores the applications of IoT in MSN and how they can address security issues. A sample case study of healthcare, discussed later in the chapter, integrates IoT with MSNs that incorporate various security mechanisms like encryption, multi-factor authentication, intrusion detection systems, firewalls, digital signatures, and access controls to ensure the confidentiality, integrity, and availability of user's data, to enhance the safety and reliability of ever growing MSNs.

5.2 MOBILE SOCIAL NETWORKS

MSNs combine social and mobile networking techniques. It enables people to share content and communicate. These systems are designed to aid in various functions, such as detecting communities and disseminating information. MSNs have become an indispensable part of our lives as a result of the widespread use of mobile devices. People can use their current social networks anywhere and anytime. This has also led to the introduction of a number of mobility-oriented applications, including location-based services, built-in cameras for taking and sharing photos and videos, and alerts via push and real-time updates.

MSN architecture consists mainly of five building blocks (Mao et al., 2017) as given in Figure 5.1.

1. The application component consists of services such as posting, searching, matching, recommendation, and other services.

2. The context mining component gathers information about the user, the device, and the environment to learn about societal characteristics, patterns of mobility, and personal preferences. Profiling and data compilation are included for developing social awareness.
3. The networking mechanisms component handles information dissemination through user addressing, neighbor discovery, and routing. It ensures proper connectivity and data delivery among users.
4. The local resource management component manages both hardware (e.g., radios, sensors, batteries, memory, and storage) and non-hardware (e.g., shareable functions) resources on users' mobile devices, supporting MSN functionality.
5. The privacy and security management component addresses anonymity, access control, and security issues. It includes schemes for user identity and location anonymity, fine-grained access control policies, and comprehensive security mechanisms against threats like eavesdropping and attacks. Overall, this architecture provides a comprehensive framework for robust and secure mobile social networks, enabling efficient application services, context mining, network connectivity, resource management, and privacy/security management.

MSN finds application in many fields and areas, like academics, medicine, business, marketing, entertainment, and so on. The MSN penetration rate has increased, which has led to a higher number of users accessing social networks. Users disclose private information, like their location, interests, and social connections, raising concerns about privacy and security (Li et al., 2011).

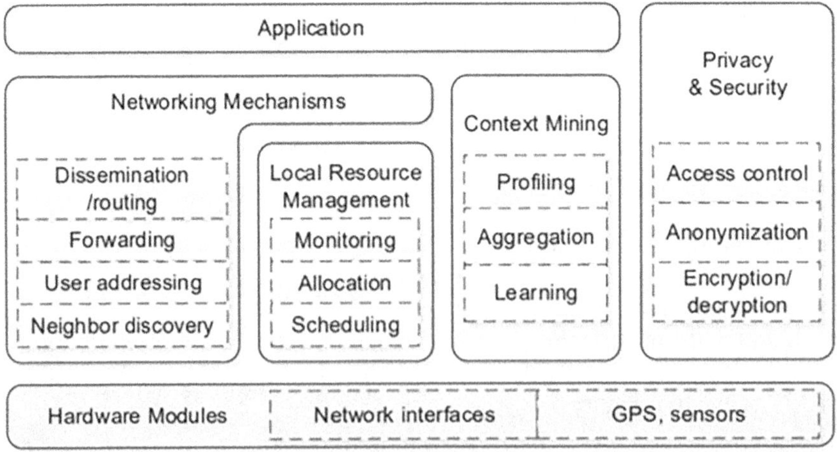

Figure 5.1 Architecture of MSN.

5.3 SECURITY AND PRIVACY ISSUES IN MOBILE SOCIAL NETWORKS

Nowadays, the most convenient and efficient way to access or find information is through a smartphone. However, because of this easy availability of information, the threat of harm arises in any normal life. Most smartphones do not come with pre-installed security software. This lack of security presents a golden opportunity for insecurity. Personal computers are equipped with a variety of security software such as firewalls and antivirus programs. Smartphones also require such security software to protect us from these threats. Nowadays, users often conduct financial transactions. They download different shopping apps and shop there and share their credit card information with them. As users of MSNs grow in significant numbers, they share information on different networks. The high accessibility and autonomy of MSNs have led to widespread rumors, scams, and other various types of misuse, which poses a grave risk to typical social network activities. This development is therefore a double-edged sword.

People have suffered significant losses as a result of the high prevalence of scams in recent years in the areas of wireless networks, online payment, and intelligent interfaces. The cost of cybercrime globally is anticipated to increase over the next five years, going from $8.6 trillion in 2022 to $23.82 trillion by 2027, according to projections from Statista's Cybersecurity Outlook as depicted in Figure 5.2 (the U.S. dollar values provided are based on the prevailing exchange rate as of November 2022).

The exponential growth in social networking sites has led to vast amounts of data and information availability on these platforms, which has increased the risk of information leakage and presents opportunities for several cybercrimes, including data interception, privacy eavesdropping, copyright infringement, and data fraud.

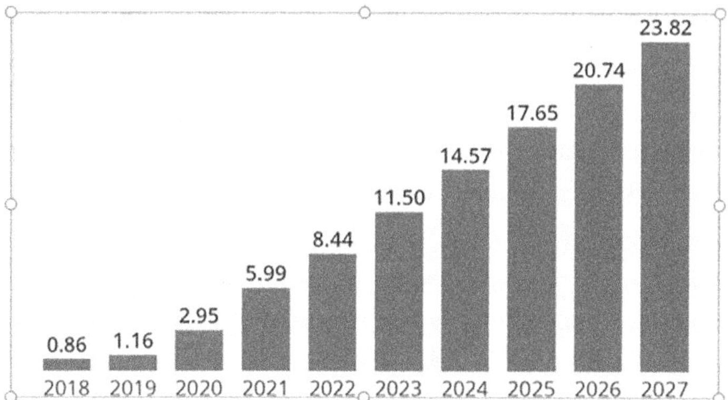

Figure 5.2 Estimated cost of cybercrime worldwide (in trillion U.S. dollars), Statista technology market outlook.

Such threats can compromise user privacy, security, and overall experience. The common categories of threats found in MSNs are the following:

- *Spam*: Unwanted mass electronic messages. MSNs can be vulnerable to malicious software called malware that can infect devices, compromise user data, and spread through shared content or links.
- *Phishing*: Specifically targeted on MSNs. It deceives users to disclose sensitive information like login passwords or personal information through malicious messages or links.
- *Identity theft*: Leads to stealing a person's identity, including their social security number, mobile phone number, and address, without their consent. Hackers get access to the victim's friend list, which they use to demand sensitive information from them using various social engineering techniques.
- *Clickjacking*: Users are tricked into clicking on hidden or misleading elements that perform unintended actions or lead to scams and fraudulent activities.
- *Account hijacking:* Unauthorized access to user accounts in MSNs that can lead to unauthorized activities, impersonation, posting of malicious content, or unauthorized access to personal information.
- *Cyber espionage*: Practice of gathering confidential information or intellectual property online with the goal of passing it to competitors.
- *Fake profiles*: MSNs suffer from the creation of fake profiles or impersonation, leading to misleading or fraudulent interactions and reputation damage.
- *Cyber bullying:* Individuals may use the platform to engage in harmful or abusive behavior towards others.
- *Cyber grooming*: During which a close, emotional bond is established with the victim, typically children and adolescents, and attackers take this liberty to threaten the child.
- *Privacy breaches*: These occur when user data is accessed, shared, or misused without consent, potentially leading to identity theft, data leakage, unauthorized profiling, or unwanted tracking of users' physical whereabouts.

It is important for users to be aware of these threats and take necessary precautions, such as using strong passwords, enabling two-factor authentication, being cautious of suspicious messages or links, regularly updating applications, and adjusting privacy settings to mitigate risks in MSNs. The privacy and security management in MSN is addressed by protecting users' identity and location, defining access control policies for personal data, and putting security measures in place to reduce threats (Mao et al., 2017). As the popularity and significance of MSNs continue to grow, it is crucial to prioritize the security and privacy of these MSN platforms.

The security requirements in MSNs to ensure that data is not breached can be summarized as follows:

- *Data integrity*: This is to ensure the accuracy and consistency of data exchanged, processed, and stored on MSNs to prevent tampering.
- *Data confidentiality*: This is required to protect the data flowing between MSN components (users and servers) from unauthorized access and untrusted entities to maintain data confidentiality.
- *Data availability*: This ensures that data is readily available at all times throughout the MSN, even in the presence of attacks.
- *Data authenticity*: Verifying the validity and authenticity of MSN components, such as users and servers, to prevent the use of invalid or forged information within the network.
- *Spam detection*: Unwanted messages or content in MSN, known as spam, can be advertisements, news, offers, or irrelevant messages. Spam can affect device performance and user experience, necessitating an accurate and efficient spam filtering system in MSNs.
- *Trust relations*: Establishing trust relationships with other entities such as users and service providers is a significant challenge in MSN. Trust is essential for secure social communication, and ensuring trustworthy interactions is crucial. It is important to make sure that identity is exchanged only after establishing trust.
- *Privacy protection*: This involves safeguarding privacy-sensitive information, including user details, device data, identity, location, and any other personal data, from unauthorized disclosure to illegal or untrusted entities within the network. This also includes safeguarding geotagging, which automatically tags the current location and exposes one's personal information like where one lives and where one is traveling, inviting thieves who can target one for robbery.
- Other than the traditional ways, like user authentication before logging on to a platform, encryption of conversations, anonymity, frequent visits to privacy settings, and geolocation restrictions, the rules for access control offering security and privacy on MSNs consist of the following:
- *Role-based access control* to limit access to data based on a user's position within the network.
- *Attribute-based access control* considers the user's characteristics, such as their location or age, to allow access.
- *Discretionary access control* (DAC): Setting permissions on posts or personal information to restrict access to only certain people or groups.
- *Mandatory access control*: Users have no choice over who has access to their information, and access is only allowed in accordance with a rigid set of criteria.

These security requirements and challenges highlight the need for robust security measures and technologies to protect user data, maintain privacy, and prevent unauthorized access within MSNs.

5.4 IoT

The IoT is a term used to describe a wide range of objects equipped with sensing and actuating devices. These devices collect, analyze, and share data among themselves, as well as with other objects, programs, and platforms. Using many physical objects and their connections facilitates the development of many smart applications and their infrastructures by IoT. All smart physical objects such as wearables, sensors, actuators, smartphones, vehicles, computers, and RFID are intelligent enough to interact with each other, and using unique addressing schemes and standard communication protocols they can cooperate with any neighboring connected smart object (Sobin, 2020).

The three main communication patterns are human to human (H2H), human to thing (H2T), and thing to thing (T2T). The IoT paradigm provides an environment where humans with smart devices (things) are capable of gathering and exchanging data over a network without human support or intervention and self-decision making. IoT architecture has three layers of operations: physical layer (sensor layer), network layer, and application layer, as depicted in Figure 5.3.

Physical layer: Physical devices with automatic sensors operate in the sensor layer, which collects information from the physical world.

Network layer: Processes the sensor data collected and passes it to the application layer, which provides services and communication to the users.

Application layer: Responsible for communication mapping between two layers.

Connecting things (including human and electronic devices) at any time and any place is possible because of MSNs. There are many applications that combine MSNs and IoT. This combination inherits social networking features and interactivity, filtration, and different services composition. This will suggest a universal framework to combine users, devices, and services and the interactions among them. This integration will bring new era of relationships that will allow the creation of novel services and applications. These will have great interest for both final users and stakeholders. It is well known that the IoT will grow by itself in the coming years and then it will be necessary to find new strategies to make the objects communicate efficiently and securely.

The majority of objects operating in IoT include sensors and communication devices enabled via wireless network. Physical devices with automatic sensors operate in the sensor layer and collect information from the physical

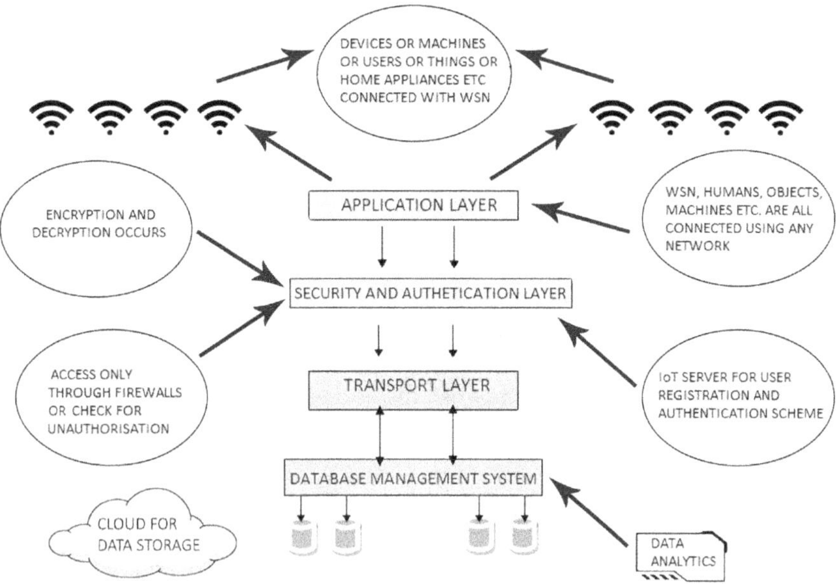

Figure 5.3 IoT architecture.

world. The network layer process the sensor data collected and passes it to the application layer, which provide services and communication to users. To ensure user privacy and data security in IoT, trust in every layer of this architecture must be implemented. Trust management in IoT aims to provide confidential services from application layer between two communication entities, including communication between two nodes or between two groups of nodes. To ensure data privacy, integrity and confidentiality must be preserved in the sensor layer.

5.5 SCOPE OF IoT IN MSN

Prominent domains where IoT has assumed pivotal roles within MSNs include the following:

- *Smart devices*: IoT-based smart devices are hardware devices such as sensors, gadgets, appliances, and other machines that collect and exchange data over the internet.
- *Home automation*: IoT home automation involves utilizing internet-connected devices to control various household appliances. It can include setting up intricate heating and lighting systems and managing home security controls and alarms, all of which are linked to the central hub and can be operated remotely using a smartphone app.

- *Wearables*: Wearable technology has many uses, including health and fitness tracking, chronic disease management, interactive gaming, performance monitoring, and navigation tracking.
- *Agriculture*: IoT smart agriculture products are designed to help monitor crop fields using sensors and by automating irrigation systems.
- *Healthcare*: Given their utility within medical facilities, IoT devices are instrumental in hospitals for monitoring patient health. They can also be equipped with sensors to track the real-time locations of medical equipment such as wheelchairs, defibrillators, nebulizers, oxygen pumps, and other monitoring devices.
- *Retail*: IoT helps brands track the location, condition, and movement of their assets and inventory anytime, anywhere. Retailers get 360-degree visibility by connecting and exchanging data with systems and devices over the internet.

5.6 ENHANCING MSN SECURITY THROUGH SIoT (SECURE IoT)

The popular IoT authentication methods for ensuring security are shown in Figure 5.4 (Zhao, 2023). It depicts various methods of IoT authentication, including password based, token based, biometric, and cryptographic authentication. Multi-factor authentication (MFA) is employed, combining traditional methods (e.g., passwords) with IoT-based authentication factors (e.g., biometrics or physical tokens). Mutual authentication between devices and servers is performed to prevent man-in-the-middle attacks. Secure device identification mechanisms, such as unique device certificates and cryptographic keys, are employed to verify the authenticity of IoT devices.

- *Behavioral biometrics and user profiling* are leveraged to identify and prevent fraudulent activities or unauthorized access. Authorization mechanisms are implemented to ensure that only authorized and trusted IoT devices can access social media accounts.

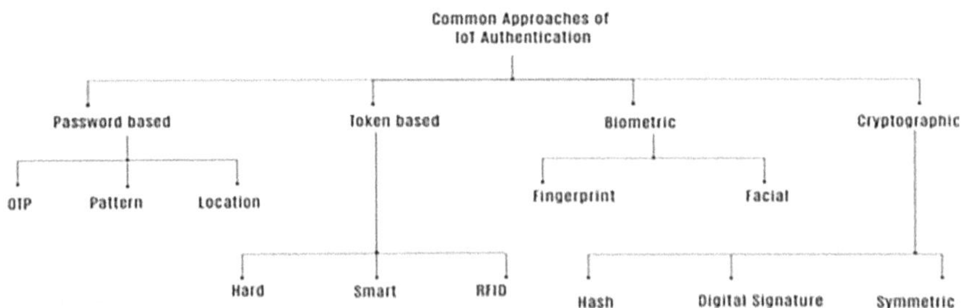

Figure 5.4 Methods of IoT authentication.

- *Secure communication channels*: Encrypted communication protocols, such as transport layer security (TLS), are used to establish secure connections between IoT devices, mobile devices, and social media servers.
- *Secure data transmission protocols*, like HTTPS, are used to protect user data during transit.
- *User identity protection*: User privacy settings are incorporated, allowing users to control the sharing and visibility of personal information or content.
- *Anomaly detection and threat monitoring*: IoT sensors and devices monitor user behavior patterns, network traffic, and device activities to detect anomalies or suspicious behavior.
- *Machine learning algorithms and AI-based analytics* are applied to identify potential security threats, such as account compromises or malicious activities.
- *Real-time alerts and notifications* are generated when suspicious activities are detected, enabling proactive response and mitigation.
- *Secure remote management*: IoT devices can provide remote management capabilities for social media accounts. Users can remotely control access permissions, revoke device access, and manage privacy settings through IoT-enabled interfaces.
- *Lost or stolen mobile devices* can be remotely locked or wiped to prevent unauthorized access to social media accounts.
- *Regular security updates and patch management*: Firmware updates and security patches for IoT devices are regularly provided by manufacturers and promptly applied to address any discovered vulnerabilities.
- *Social media platforms and mobile applications* implement regular security updates to address potential security risks and maintain compatibility with evolving IoT security standards.

It is important to note that the implementation of such a model would require collaboration between IoT device manufacturers, social media platform providers, and mobile application developers. Additionally, thorough security testing, adherence to industry standards, and continuous monitoring of the IoT ecosystem are crucial to ensuring a robust security posture for mobile social media platforms.

An extended version of IoT called Secure IoT (SIoT) has emerged as a paradigm for providing more secure and efficient information delivery through MSNs. Because security and privacy are major concerns in MSNs, SIoT focuses primarily on user privacy, safe communications, and trustworthy interactions. These are the most sensitive requirements for the success of MSN and IoT integration. By incorporating secure technology through the SIoT paradigm, the integration of IoT and MSN can establish itself as a robust technological framework. Methods such as data confidentiality and user privacy, effective ID management, and privacy-enhancing technologies represent the basis that will allow users to trust the SIoT ecosystem (Ortiz et al., 2014).

Figure 5.5 presents a secure IoT ecosystem in which all connected components use any wireless sensor network (WSN)—all can communicate at the application layer. When they first try to connect with any device, access will be authenticated to prevent the threat of unauthorized access. Data will be transferred through an encryption and decryption technique, so that if hacking occurs in the transit of data, the data cannot be leaked or understood by hackers. In other words, encryption helps to protect data from being compromised. It protects data that is being transferred as well as data stored. Although encryption helps to protect data from any unauthorized access, it does not prevent data loss. The security and authentication layer is responsible for managing all security and authentication matters. The IoT server would also work in the collaboration of these layers. After the security and before the storage layer, there would be a transport layer. This is responsible for routing and managing the network and secures the appropriate route for data transfer, so that data can travel in an efficient and safe manner. Last is the storage layer, which stores data and connects it to the devices. When considering social networks and professionals' data, big data management would be necessary. For this, bulk storage s required to ensure that no hurdles are encountered in storing and retrieving data. In this architecture, data storage would be managed on the cloud. When all data are on cloud, any user, device, or thing can retrieve or access data from anywhere and at any time.

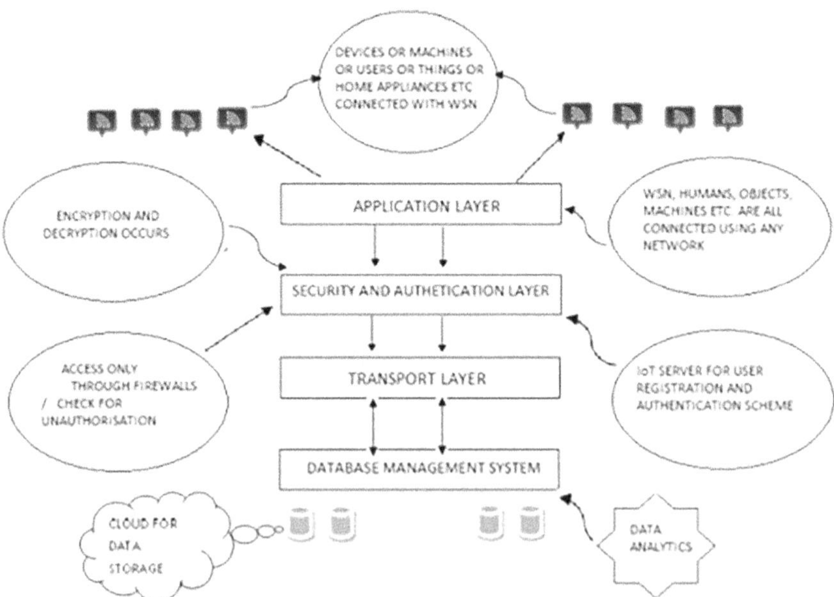

Figure 5.5 Layered architecture of the devices connected with any network.

Anywhere, anytime data availability makes this architecture more flexible and versatile. Anytime connection could also be accessed by a web services API, which would also have administrator access control over data. So this is again an authentication check for accessing stored data. There are other key mechanisms like access control, auditing, authentication, and authorization for protecting data. In the cloud, if data would be lost for any reason, then a backup option would also be there. In addition to the advantages offered by cloud usage, it's important to address several challenges, including interoperability, reliability, availability, security, privacy, and portability.

5.7 CASE STUDY: MSN IN HEALTHCARE

The presence of mobile devices in the healthcare environment has led to rapid growth in the development of medical-based applications that assist in patient care. Many apps assist healthcare professionals (HCPs) in such important tasks as information and time management, health record maintenance and access, communications and consulting, reference and information gathering, patient management and monitoring, clinical decision-making, and medical education and training.

Such apps provide many benefits for HCPs, including increased access to point-of-care tools shown to support better clinical decision-making and improved patient outcomes. However, some HCPs remain reluctant to adopt the use of such tools. Despite the benefits they offer, better standards and validation practices regarding mobile medical apps need to be established to ensure the proper use and integration of these increasingly sophisticated tools into medical practice.

IoT healthcare applications assist doctors in the prevention, practice, and diagnosis of medical conditions. Continuous automated monitoring and better patient condition analysis result in precise, error-free treatment. IoT healthcare devices collect highly reliable data that allow doctors to make informed decisions.

Proper usage of IoT can help to resolve various medical challenges like speed, price, and complexity. Use of IoT-based healthcare devices increased during the COVID pandemic, as they could be easily customized to monitor caloric intake and treatment of asthma, diabetes, and arthritis of the COVID-19 patient. This digitally controlled health management system can improve the overall performance of healthcare during pandemic days. IoT has a favorable effect on healthcare, enhancing millions of people's lives. It extensively examines the healthcare system and aids in the detection of disease. It gives each person individualized attention to their advantage. IoT technologies can provide a variety of information, including appointment reminders, blood pressure readings, calorie counts, disease status, and much more. A system like this is getting more popular because it reduces patient and doctor visits.

5.8 CREATION OF SECURE MSN USING IoT IN HEALTHCARE

A healthcare system developed for remote patient data diagnosis and analysis must ensure safe and continuous data flow. A suggested system, shown in Figure 5.6 (Refaee et al., 2022), for creating secure MSN using IoT in healthcare involves analyzing the dataset for IoT sensors, preprocessing the data with data cleaning and data reduction, feature extraction with fuzzy dynamic trust-based Routing Protocol for Low Power and Lossy Networks (RPL) routing protocol, and a butter ant optimization (BAO) algorithm based on secure and scalable healthcare data transmission in IoT for mobile computing applications. Steps are discussed in detail as follows:

- *Data collection*: The dataset includes recordings of participants' bodily movements and vital signs as they perform different physical exercises. The trained motions of the patient are observed using sensors on the ankle, wrist, and chest.
- *Preprocessing*: During this phase network traffic is transformed into a sequence of observations, with each piece of data expressed as a feature vector through preprocessing. It involves data cleaning and data reduction.

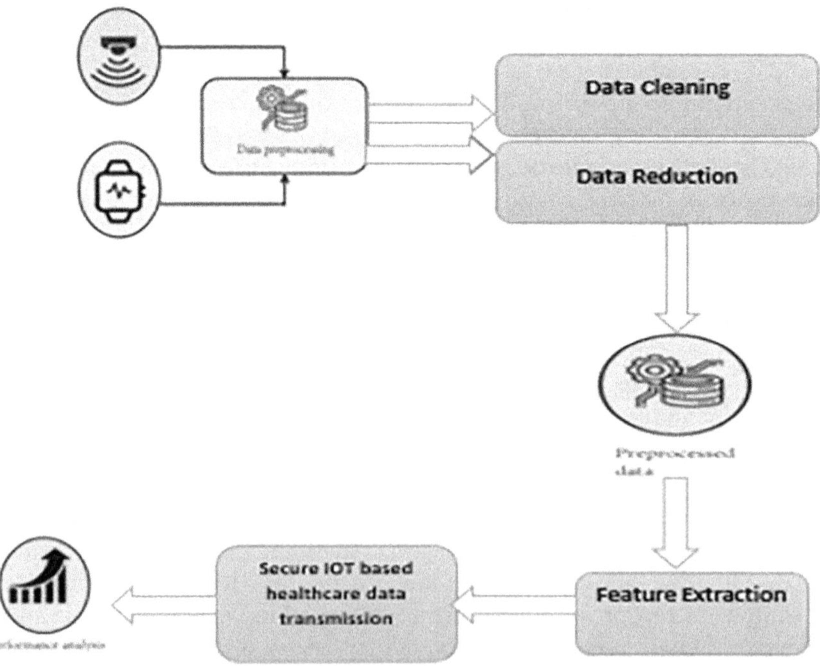

Figure 5.6 Creation of secure MSN using IoT for healthcare industry.

- *Data cleaning*: This stage's main goal is to analyze the data and discover its distinct properties. This stage includes locating errors, missing values, and corrupted records. K-nearest neighbor technique, a powerful machine learning algorithm for categorizing and identifying patterns, is also utilized in this stage.
- *Data reduction*: Data reduction techniques must be used when dealing with high-dimensional data since there are many dimensions that are extraneous, which can hide existing clusters in noisy data, as well as because the processing would become more complex and might compromise real-time requirements. The principal component analysis (PCA) is used in the model.

The preprocessed data used to extract the features employs modified local binary patterns (MLBP). This data is given as input to fuzzy dynamic trust-based RPL routing protocol combined with the BAO algorithm for low-power and loss networks, the proposed fuzzy dynamic trust-based RPL (FDT-RPL) protocol improves the overall security of data transmission.

The algorithm has been implemented for a smart healthcare system, and its performance was found to be better than conventional methods (Refaee et al., 2022). The proposed routing protocol offered a scalable and secure method of transmitting medical data.

5.9 ADVANTAGES AND LIMITATIONS OF IoT-BASED MSN

The IoT-based network collects and transmits data without the need of human interaction over a wireless network. Smart wearables, smart health monitoring, traffic monitoring, IoT in agriculture with many sensors, smart devices, hospital robotics, smart grid and water supply, and other IoT-based applications are just a few examples of how it is being used in daily life. There are advantages and disadvantages to implementing IoT in MSNs.

5.10 ADVANTAGES

- It can aid in the smarter control of homes and cities through mobile phones and save a lot of time by automating activities.
- Information is readily available and frequently updated.
- The efficient use of electricity is possible due to the direct connection and communication of electric devices with a controller computer.
- IoT apps can provide personal support by sending reminders and notifications of your usual plans.
- IoT devices connect and interact with one another and execute a variety of functions without requiring human intervention, which thus minimizes human work.

- Patient care is provided more effectively in real time without the need for a doctor's visit. It gives patients the ability to make choices as well as provide evidence-based care.
- IoT has made inventory control, traffic tracking, supply chain management, delivery, surveillance, individual order tracking, and customer management more cost-effective.

5.11 DISADVANTAGES

- Hackers might get into the system and take personal data. There is a risk of misuse of personal information because so many devices are connected to the internet.
- They are largely reliant on the internet and cannot function properly without it.
- We become less in charge of our life. Overusing the internet and other technological advancements makes people less intellectual because they become lazy and completely dependent on their smart devices rather than performing physical labor.
- Deploying IoT devices is very expensive and time consuming. Planning, building, managing, and enabling a broad technology to IoT framework are quite challenging.

5.12 CONCLUSION

Mobile social networks allow users to connect, communicate, and interact with each other through social media platforms such as Facebook, Twitter, and so on. They have transformed how individuals engage and communicate with each other by enabling users to instantly express their ideas, emotions, and experiences with each other. Nowadays, the most convenient and efficient way to access or find information is through a smartphone. However, because of this easy availability of information, the threat of harm to information that we share on these platforms has increased. Most smartphones do not come with pre-installed security software. According to the official data released by the security ministry, telecommunications fraud in MSNs has grown at an annual rate of 20%–30%. Threats such as spams, identity theft, and account hijacking can compromise user privacy, security, and overall experience with the social media platform. The use of IoT in these MSNs for sharing information has further increased security risks. This is where secure IoT comes into play, focusing on developing a secure IoT-based MSN. IoT has benefits but also disadvantages, which should be kept in mind while implementing IoT in our system.

REFERENCES

A. M. Ortiz, D. Hussein, S. Park, S. N. Han, and N. Crespi, "The cluster between internet of things and social networks: Review and research challenges," *IEEE Internet of Things Journal*, vol. 1, no. 3, pp. 206–215, June 2014. https://doi.org/10.1109/JIOT.2014.2318835.

C. Sobin, "A survey on architecture, protocols and challenges in IoT," *Wireless Personal Communications*, vol. 112, pp. 1383–1429, January 2020.

Eshrag Refaee et al., "Secure and scalable healthcare data transmission in IoT based on optimized routing protocols for mobile computing applications," *Wireless Communications and Mobile Computing*, vol. 2022, pp. 1–12, 2022.

H. H. Junhui Zhao, "Authentication technology in internet of things and privacy security issues in typical application scenarios," *Electronics*, vol. 2023, 2023.

M. Li, N. Cao, S. Yu, and W. Lou, "Findu: Privacy-preserving personal profile matching in mobile social networks," *2011 Proceedings IEEE INFOCOM*, IEEE, April 2011, pp. 2435–2443.

Z. Mao, Y. Jiang, G. Min, S. Leng, X. Jin, and K. Yang, "Mobile social networks: Design requirements, architecture, and state-of-the-art technology," *Computer Communications*, vol. 100, pp. 1–19, 2017.

Chapter 6

End-User and Human-Centric IoT, Including IoT Multimedia, Societal Impacts and Sustainable Development

Garima Tyagi

6.1 INTRODUCTION

The Internet of Things (IoT) is a rapidly growing technology that has the potential to revolutionize many aspects of our lives, from smart homes and cities to connected healthcare and transportation. However, as the number of IoT devices and systems continues to grow, it is increasingly important to consider the needs and preferences of the end-users, as well as the impact of these technologies on society and the environment. End-user and human-centric IoT is an approach to designing and developing IoT systems that focuses on these factors, with the goal of creating technology that is not only functional and efficient, but also socially and environmentally responsible. This approach includes the integration of multimedia functionality into IoT devices and systems, such as the ability to capture and transmit audio and video, in order to enhance the user experience and enable new applications. Additionally, it also considers the societal and environmental impacts of IoT deployment, such as privacy concerns, energy consumption and the role of IoT in sustainable development. In this chapter, we will explore the concept of end-user and human-centric IoT, including IoT multimedia, societal impacts and sustainable development.

6.2 INTERNET OF THINGS

The Internet of Things, commonly referred to as IoT, is a revolutionary concept that has been rapidly transforming the way we interact with the world around us. It involves connecting everyday physical objects and devices to the internet, enabling them to collect and exchange data, communicate with each other, and perform various tasks autonomously. The core idea behind IoT is to create a seamlessly interconnected network of devices, systems, and services, resulting in smarter, more efficient, and data-driven experiences.

The Internet of Things refers to the interconnectivity of everyday devices and appliances, allowing them to communicate and share data with each

 DOI: 10.1201/9781003406723-6

other over the internet. This technology has the potential to revolutionize many industries and aspects of our daily lives, from smart homes and cities to connected healthcare and transportation. IoT devices can range from simple sensors and actuators to more complex devices such as smartphones, laptops, and smart appliances. These devices collect, share and act on data, enabling new forms of automation, optimization and predictive maintenance. This technology creates a more connected world, where devices can communicate with each other, enabling them to work together to make our lives more convenient, efficient, and safe. As the number of IoT devices continues to grow, it is increasingly important to consider the potential impacts of this technology on privacy, security, and society as a whole.

The term "Internet of Things" encompasses a broad vision where everyday objects, places, and environments are interconnected through the internet. An example of an IoT object found in some homes is a thermostat that can determine when a person occupies specific rooms, adjusting heating, lighting, and other functions accordingly. Expanding the internet from "a network of interconnected computers to a network of interconnected objects," the IoT involves a diverse network of interconnected devices. These devices include various sensors to measure their environment, actuators to physically interact with their surroundings, nodes to relay information, and coordinators to manage sets of these components.

This interconnected network has the potential to significantly impact and enhance the relationship between individuals and their surroundings. Many believe that the IoT will address societal challenges such as an aging population, deforestation, and recyclability.

The interconnection of physical objects is expected to amplify the profound effects of large-scale networked communications on society. The IoT envisions a world where everyday objects are connected to share and utilize information, emphasizing the role of people alongside inanimate objects. While smartphones are commonplace today, the IoT extends beyond them. Future sensors measuring health and movement within our living environment will help us navigate and understand the world in ways we can hardly imagine.

It is crucial for researchers, software engineers, and entrepreneurs driving IoT development to implement the technology responsibly. Policymakers can support responsible innovation and decide whether and how to regulate if necessary. All stakeholders in the IoT should continuously anticipate potential outcomes rather than simply react to unforeseen consequences.

Human-computer interface (HCI) has engaged with technologies sharing characteristics with those proposed for the IoT, particularly in ubiquitous and pervasive computing efforts that utilize internet technologies for sensing, tracking, and monitoring.

The core vision of ubiquitous and pervasive computing aligns with that of the IoT, aiming to integrate computers into everyday objects, devices, and displays. Wearable computing, focusing on embedding computers into ordinary items, shares attributes with the IoT, with the primary distinction being the emphasis on interconnectivity (Berzowska, 2005).

Efforts in pervasive, ubiquitous, and wearable computing have typically involved a single device connecting to a corresponding data source (Berzowska, 2005). In contrast, the IoT introduces the concept of an ecosystem where one device communicates with many interconnected things.

This chapter reviews the latest HCI-related literature and commercial products related to the IoT vision. The aim is to provide a resource for the general HCI audience to understand the current state of research related to the emerging IoT initiative. Another goal is to highlight ways in which HCI can more effectively engage with IoT efforts by fostering a dialogue between literature findings and industry offerings, revealing opportunities and future approaches for HCI in full integration with the IoT. The chapter then describes the methodology used in the survey, presents the findings, and discusses how the HCI community can engage with the evolving IoT vision, supported by tables classifying recent research and business efforts related to the IoT.

6.3 KEY COMPONENTS OF IoT

1. *Devices and Sensors*: IoT relies on a vast array of devices and sensors that can collect data from the environment and perform specific actions. These devices can range from simple temperature and humidity sensors to complex devices like smart thermostats, wearables, industrial machines, and even autonomous vehicles.

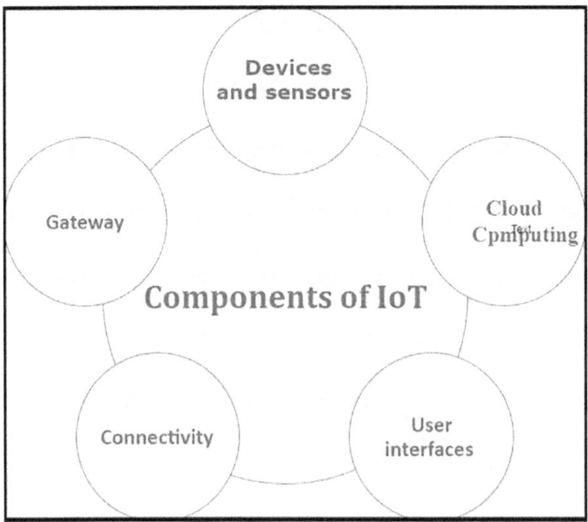

Figure 6.1 Components of IoT.

2. *Connectivity*: For IoT to function effectively, the connected devices need to communicate with each other and share data. This is made possible through various communication protocols such as Wi-Fi, Bluetooth, Zigbee, cellular networks, and more. The choice of connectivity depends on the specific requirements of the IoT application.
3. *Data Processing and Analytics*: IoT generates an enormous amount of data from interconnected devices. To make this data meaningful and actionable, it needs to be processed and analyzed in real time. Advanced data analytics, cloud computing, and edge computing technologies play a vital role in extracting valuable insights from the collected data.
4. *Cloud Computing*: Cloud platforms are essential components of IoT systems. They provide the necessary infrastructure to store, process, and analyze data from IoT devices. Cloud services also enable seamless access to data and applications from anywhere and at any time.
5. *Security and Privacy*: With the increasing number of connected devices, ensuring the security and privacy of data becomes a critical challenge. IoT systems must employ robust security measures to safeguard against cyber threats and unauthorized access.

6.4 APPLICATIONS OF IoT

IoT has found applications across various industries and sectors, some of which include the following:

1. *Smart Homes*: IoT-enabled smart homes allow homeowners to control lighting, heating, air conditioning, security systems, and appliances remotely through smartphones or voice commands.
2. *Healthcare*: IoT is transforming healthcare by enabling remote patient monitoring, wearable health devices, smart medical equipment, and efficient health management systems.
3. *Industrial IoT (IIoT)*: IIoT optimizes industrial processes by using sensors and data analytics to improve productivity, reduce downtime, and enhance overall efficiency in manufacturing and logistics.
4. *Smart Cities*: IoT technologies are being used to create smart city solutions that enhance urban services, traffic management, waste management, and energy efficiency.
5. *Agriculture*: IoT enables precision farming by providing real-time data on soil conditions, weather, and crop health, leading to optimized resource usage and increased crop yields.
6. *Transportation*: IoT applications in transportation include smart traffic management, connected vehicles, and logistics tracking for better fleet management.

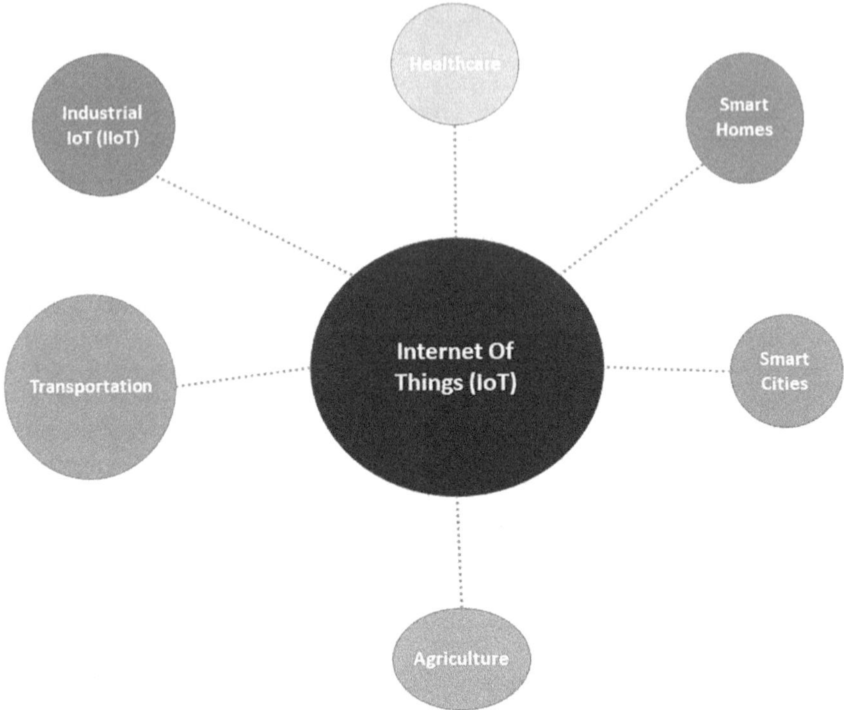

Figure 6.2 Applications of IoT.

6.5 CHALLENGES AND FUTURE OF IoT

Despite its immense potential, IoT faces several challenges, including the following:

1. *Security Concerns*: The vast network of interconnected devices presents security vulnerabilities, making IoT susceptible to cyber-attacks and data breaches.
2. *Data Privacy*: The massive amount of data collected by IoT devices raises concerns about user privacy and data ownership.
3. *Interoperability*: The lack of standardization in IoT devices and platforms can hinder seamless communication and integration.
4. *Scalability*: As the number of connected devices grows exponentially, IoT systems must be scalable to handle the increasing data load.

The future of IoT is promising, as ongoing advancements in technology and the increasing adoption of IoT solutions are likely to drive innovation and create new possibilities across various domains. However, addressing the challenges and ensuring a secure and privacy-focused approach will be crucial for its sustainable growth and success.

6.6 HUMAN COMPUTER INTERFACE

A HCI is the point of interaction between a human and a computer system. The goal of HCI is to design and develop interfaces that are easy to use, efficient, and effective for the human user. This includes both the physical interface (such as a keyboard, mouse, or touchscreen) and the software interface (such as graphical user interfaces or command-line interfaces). HCI is a multidisciplinary field that draws on research and knowledge from computer science, psychology, sociology, and other disciplines to understand how people interact with technology and to design interfaces that meet their needs.

HCI has evolved over the years with the advancement in technology, from the early days of command-line interfaces to the current trend of natural user interfaces such as voice and gesture recognition. With the increasing prevalence of IoT devices, HCI is becoming more important than ever, as it is crucial for these devices to be easy to use and to provide a seamless experience for the user. As the technology continues to advance, HCI will play an important role in shaping the future of HCI.

HCI is the study of how people interact with computers and technology. It encompasses the design, evaluation, and implementation of interactive systems, including the hardware and software that make up those systems, as well as the human factors involved in the interactions. HCI research and design focus on creating systems that are easy to use, efficient, and effective for the users and that meet their needs and goals.

The field of HCI covers a wide range of topics, including user-centered design, user experience (UX) design, human factors engineering, and usability engineering. These areas are essential to create user-friendly technology, which is easy to understand and use, and provides a seamless experience for the users. This includes designing interfaces that are intuitive, visually pleasing, and accessible, as well as providing feedback and error messages that are easy to understand.

HCI also includes the study of new interaction techniques and technologies such as natural language processing, virtual and augmented reality, and touch and gesture-based interfaces. With the rise of new technologies, the field of HCI continues to evolve and adapt to new ways of interacting with computers, devices, and information.

Here it refers to the point of interaction and communication between humans and digital devices or computer systems. It encompasses all the technologies, methodologies, and designs that enable users to interact with computers and other digital devices in a natural and intuitive manner.

6.7 KEY COMPONENTS OF HUMAN-COMPUTER INTERFACE

1. *Input Devices*: Input devices are used to provide information or data to the computer system. Common input devices include keyboards,

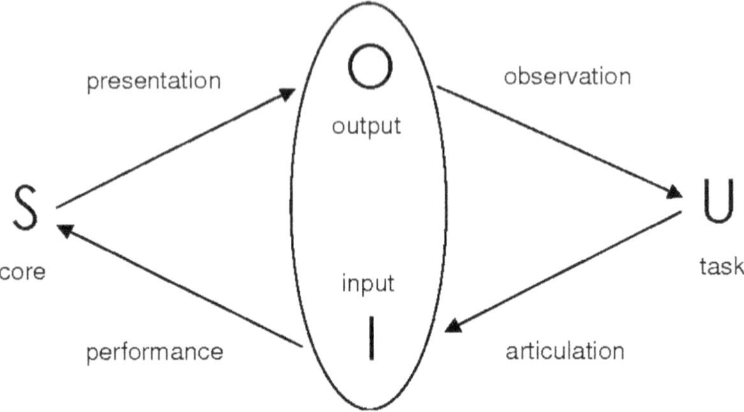

Figure 6.3 Key elements of human computer interface.

mice, touchscreens, voice recognition systems, gesture recognition devices, and more. These devices enable users to convey their intentions to the computer.

2. *Output Devices*: Output devices display the processed information back to the user. Monitors, speakers, printers, and haptic feedback devices (e.g., vibration motors) are examples of output devices that help users perceive and interpret the computer's responses.
3. *User Interface (UI)*: The user interface is the visual and interactive design that allows users to interact with the software or application. A well-designed UI aims to be intuitive, user-friendly, and responsive, ensuring a seamless and satisfying user experience.
4. *Interaction Techniques*: Interaction techniques refer to the methods used to interact with the computer system. These may include clicking, dragging, typing, voice commands, multi-touch gestures, and other modalities designed to cater to user preferences and tasks.
5. *Feedback Mechanisms*: Providing feedback is vital to HCI. Users need to know that their actions have been recognized and that the system is responding appropriately. Visual cues, auditory alerts, and haptic feedback are some examples of feedback mechanisms.
6. *Accessibility*: HCI should consider the needs of all users, including those with disabilities. Accessible interfaces ensure that everyone can interact with technology effectively, regardless of their abilities.

6.8 DESIGN PRINCIPLES OF HUMAN-COMPUTER INTERFACE

1. *User-Centered Design (UCD)*: User-centered design places the user at the center of the design process. It involves understanding the users'

goals, needs, and tasks through methods like user interviews, surveys, and observation. By gaining insights into user behavior and preferences, designers can create interfaces that are tailored to meet users' expectations and facilitate their tasks. UCD promotes empathy, as designers strive to see the interface from the user's perspective, resulting in more intuitive and user-friendly designs.

2. *Consistency*: Consistency is one of the fundamental principles of HCI design. It ensures that elements, interactions, and visual cues are uniform throughout the interface and across different applications within a system. Consistent design patterns help users build mental models, enabling them to predict how interactions will work in various contexts. Consistency simplifies the learning process and reduces the cognitive load on users, making the interface more efficient and enjoyable to use.
3. *Clarity and Simplicity*: The principle of clarity and simplicity emphasizes presenting information, actions, and content in a straightforward and understandable manner. This involves using clear language, simple icons, and logical organization to avoid confusion. By removing unnecessary complexities, designers can enhance the usability of the interface and ensure that users can accomplish their tasks with ease.
4. *Feedback and Affordance*: Providing timely and appropriate feedback is essential for effective HCI. Users should receive immediate responses when they interact with the interface, confirming that the system has recognized their inputs. Affordance is about designing elements in a way that suggests their intended functionality. For example, a button should visually appear as clickable, indicating to users that it can be pressed. Feedback and affordance help users understand the state of the system and guide their actions.
5. *Learnability*: Learnability refers to the ease with which users can learn how to use the interface effectively. A well-designed HCI should be intuitive enough for users to quickly understand how to interact with it without requiring extensive training. Designers achieve learnability by employing familiar design patterns, adhering to industry standards, and providing clear signifiers for interactive elements.
6. *Flexibility*: HCI should be flexible enough to accommodate different user preferences, workflows, and contexts. Offering customization options, such as configurable layouts or settings, empowers users to tailor the interface to their specific needs. Designing for flexibility is particularly important in systems that cater to diverse user groups with varying levels of expertise and requirements.
7. *Error Prevention and Handling*: An effective HCI design takes into account potential user errors and incorporates measures to prevent them or provide guidance for recovery. Error prevention can include confirmation dialogs for critical actions or designing interfaces that prevent users from making irreversible mistakes. When errors do

occur, clear and informative error messages help users understand the problem and take appropriate corrective actions.

8. *Visibility and Accessibility*: Important information and key features should be easily visible and accessible to users. Organizing the interface in a logical and hierarchical manner helps users find what they need without unnecessary searching. Accessibility is a crucial aspect of HCI design, ensuring that the interface is usable by people with disabilities. Designers must consider inclusive practices, such as providing alternative text for images and keyboard navigation for users who cannot use a mouse.
9. *Minimalism*: The principle of minimalism advocates for simplicity and removing unnecessary elements from the interface. Striving for a clean and uncluttered design reduces distractions and enhances the user's focus on essential functionalities. Minimalist interfaces can lead to improved user productivity and engagement.
10. *Aesthetic Integrity*: Aesthetic integrity concerns the consistency and coherence of the visual design elements throughout the interface. By adhering to a cohesive design language, including color schemes, typography, and iconography, designers create a visually pleasing and harmonious user experience. Aesthetically appealing interfaces can contribute to a positive emotional response and user satisfaction.
11. *Progressive Disclosure*: In interfaces with complex functionalities, presenting all options at once can overwhelm users. Progressive disclosure involves revealing information or advanced features gradually, based on the user's needs and actions. Designers can use techniques like collapsible menus, tooltips, or contextual help to simplify the initial interaction and provide additional functionalities when required.
12. *Context Sensitivity*: Context sensitivity implies that the interface adapts to the user's current context or task. By providing relevant options and information based on the user's actions or current state, designers can streamline the user experience and eliminate unnecessary steps. Context-sensitive interfaces cater to the user's specific needs, promoting efficiency and reducing cognitive overhead.

6.9 ADVANCEMENTS IN HUMAN-COMPUTER INTERFACE

1. *Natural Language Processing (NLP) and Voice Interfaces*: NLP technology enables computers to understand and interpret human language. Voice interfaces, powered by NLP, allow users to interact with devices through natural language commands and spoken words. This advancement has led to the widespread adoption of voice assistants like Amazon Alexa, Apple's Siri, Google Assistant, and others. Voice interfaces are becoming more accurate and capable, enabling

users to perform various tasks, such as setting reminders, controlling smart home devices, and finding information, using natural voice interactions.

2. *Gesture Recognition*: Gesture recognition technology allows users to interact with computers and devices through hand movements and gestures. Advanced cameras, depth sensors, and machine learning algorithms enable devices to recognize and interpret gestures accurately. This advancement has paved the way for intuitive and immersive interactions, especially in virtual reality and augmented reality applications, where users can manipulate virtual objects using natural hand movements.
3. *Virtual Reality (VR) and Augmented Reality (AR)*: VR and AR technologies have significantly impacted HCI, particularly in gaming, training, education, and visualization applications. VR immerses users in a simulated digital environment, while AR overlays digital content onto the physical world. These technologies create new opportunities for interactive experiences, such as virtual tours, immersive training simulations, and interactive gaming, enhancing user engagement and learning outcomes.
4. *Brain-Computer Interfaces (BCI)*: BCI technology establishes a direct communication channel between the human brain and computers or external devices. Electroencephalography (EEG) and other brainwave-sensing techniques allow users to control devices or perform tasks using their thoughts alone. BCI holds great potential for assisting people with severe physical disabilities, enabling them to interact with technology, communicate, and control their environment through brain signals.
5. *Tangible User Interfaces* (TUI): TUIs incorporate physical objects and manipulatives into the interaction with digital systems. By combining the benefits of physical and digital interactions, TUIs create more tangible and immersive user experiences. For example, children can learn coding concepts by using physical blocks representing commands to control digital characters in educational games.
6. *Eye Tracking*: Eye tracking technology allows computers to monitor the movement and gaze of a user's eyes. By understanding where the user is looking, HCI systems can adapt the interface, offer gaze-based interactions, and provide insights into user behavior and attention. Eye tracking is being used in diverse applications, including assistive technologies, market research, and gaming.
7. *Emotion Recognition*: Emotion recognition technology aims to identify and interpret human emotions through facial expressions, vocal cues, or physiological signals. Integrating emotion recognition into HCI enables more personalized and adaptive experiences, as systems can respond appropriately based on the user's emotional state. This can be particularly valuable in areas like mental health, virtual therapy, and customer service.

8. *Brainwave Sonification*: Brainwave sonification is a unique approach that converts brainwave activity into sound, allowing users to listen to their brain's electrical activity in real time. This technique can help users understand their mental state and engage in activities like meditation or relaxation.
9. *Biofeedback Devices*: Biofeedback devices provide users with real-time information about their physiological or mental states, such as heart rate, stress levels, and brainwave patterns. These devices promote self-awareness and mindfulness and are used in various fields, including stress management, sports training, and healthcare.
10. *Facial Recognition and Emotion Synthesis*: Facial recognition technology has advanced to accurately identify individuals based on their facial features. Emotion synthesis, on the other hand, allows devices to generate expressive and emotionally responsive avatars. These technologies find applications in virtual meetings, teleconferencing, and gaming, providing more engaging and human-like interactions.

Advancements in HCI continue to push the boundaries of technology and human interaction. As these technologies mature and become more accessible, they hold the potential to revolutionize various industries and further enhance the way we interact with computers and digital systems in the future. As a result, the boundaries between the digital and physical worlds are becoming increasingly seamless, opening up exciting possibilities for various fields, including education, healthcare, entertainment, and productivity.

6.10 HUMAN-CENTRIC IoT: ENHANCING USER EXPERIENCE IN THE INTERNET OF THINGS

Human-centric IoT, often referred to as human-centered IoT or HCIoT, is an approach to designing and implementing IoT systems with a strong focus on enhancing the overall user experience. It recognizes that the success of IoT applications lies not only in their technical capabilities but also in how effectively they address human needs, preferences, and aspirations. By placing human users at the core of IoT development, HCIoT aims to create seamless, intuitive, and meaningful interactions between humans and connected devices, thereby maximizing the potential of IoT technology in various domains.

6.11 KEY ASPECTS OF HUMAN-CENTRIC IoT

1. *User-Centered Design*: At the heart of HCIoT is the principle of UCD. This approach involves understanding users' behaviors, needs, and pain points through comprehensive user research. By involving

end-users in the design process from the early stages, HCIoT projects can create IoT solutions that are tailored to their expectations, resulting in increased adoption and user satisfaction.

2. *Context Awareness*: HCIoT systems take into account the context in which users interact with IoT devices. Context awareness means that IoT applications can adapt and respond to changes in the user's environment, preferences, and behaviors. By being contextually aware, these systems can deliver personalized and relevant experiences, making them more useful and valuable to users.
3. *Simplicity and Intuitiveness*: HCIoT prioritizes simplicity and intuitiveness in design. It aims to reduce the complexity of interactions and minimize the cognitive load on users. IoT interfaces should be easy to understand and use, even for non-technical users. This emphasis on simplicity helps avoid user frustration and encourages broader adoption of IoT technology.
4. *Emotion-Aware Computing*: Human emotions play a significant role in how users perceive and engage with technology. HCIoT seeks to understand and respond to user emotions, creating emotionally aware computing systems. For example, an emotion-aware home automation system might adjust lighting and temperature settings based on the user's mood.
5. *Privacy and Security*: HCIoT recognizes the paramount importance of privacy and security in IoT deployments. By safeguarding user data and ensuring secure communication between devices, HCIoT builds trust and confidence among users, encouraging them to embrace IoT solutions.
6. *Inclusivity and Accessibility*: HCIoT aims to be inclusive, considering the needs of all users, including those with disabilities. Ensuring accessibility features in IoT interfaces makes them usable and valuable to a wider range of people, promoting equality and diversity in IoT adoption.

6.12 EXAMPLES OF HUMAN-CENTRIC IoT APPLICATIONS

1. *Smart Agriculture*: In the context of smart agriculture, HCIoT facilitates precision farming, where sensors collect data on soil conditions, weather, and crop health. This data is then used to optimize irrigation, fertilization, and pest control, resulting in increased crop yields and reduced resource wastage. HCIoT interfaces empower farmers with actionable insights and intuitive controls to manage their agricultural operations effectively.
2. *Connected Healthcare Devices*: HCIoT is driving the proliferation of connected healthcare devices, such as smartwatches, fitness trackers,

and health monitors. These devices continuously track vital signs, activity levels, and sleep patterns, enabling users to proactively manage their health. HCIoT applications provide user-friendly mobile apps that display personalized health data and offer health recommendations, encouraging users to adopt healthier lifestyles.

3. *Environmental Monitoring and Conservation*: HCIoT plays a crucial role in environmental monitoring and conservation efforts. Sensors deployed in natural habitats collect data on air and water quality, wildlife behavior, and climate conditions. HCIoT applications provide researchers and conservationists with real-time data visualizations and analysis tools, aiding in biodiversity conservation and environmental protection.
4. *Smart Energy Management*: In the realm of smart energy management, HCIoT empowers users to monitor and control their energy consumption efficiently. Smart meters and home energy management systems provide real-time data on energy usage, enabling users to identify energy-saving opportunities. User-friendly interfaces and personalized energy reports encourage energy conservation and cost savings.
5. *Personalized Learning and Education*: HCIoT contributes to personalized learning experiences in education. Smart educational tools and platforms adapt content and learning materials based on individual learning styles and progress. By collecting data on student performance and preferences, HCIoT applications offer personalized recommendations and learning pathways for students.

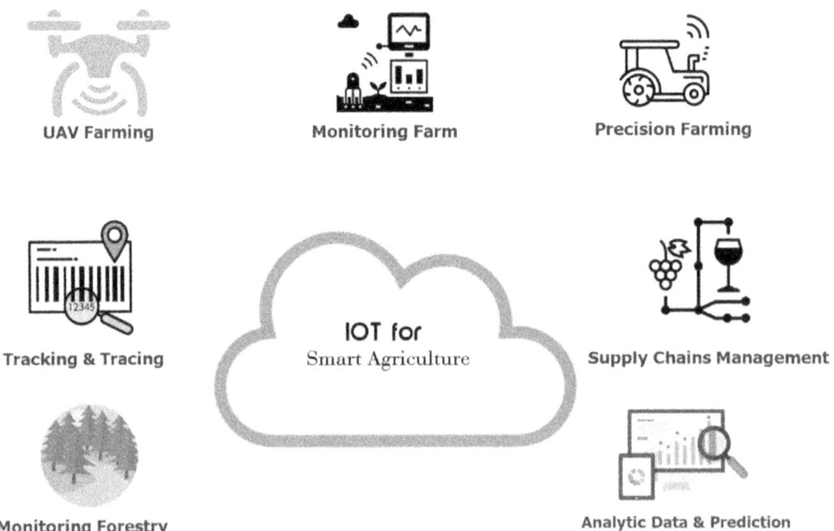

Figure 6.4 Smart agriculture.

6. *Smart Cities and Urban Planning*: HCIoT drives the development of smart city solutions, making urban living more efficient and sustainable. IoT-enabled sensors and infrastructure monitor traffic flow, waste management, air quality, and public services. HCIoT interfaces provide citizens with real-time information and services to enhance their urban experiences and promote community engagement.
7. *Home Healthcare and Remote Monitoring*: HCIoT is transforming home healthcare by enabling remote patient monitoring. IoT devices track patients' health metrics and share data with healthcare providers in real time. User-friendly apps and interfaces help patients and caregivers manage health conditions effectively from the comfort of their homes.
8. *Social IoT for Community Engagement*: HCIoT fosters social IoT applications that encourage community engagement and social interactions. For example, IoT-powered community hubs can provide information on local events, facilitate neighborhood communication, and encourage shared resources among residents.
9. *Elderly Care and Aging in Place*: HCIoT applications support aging in place by providing smart home solutions that enhance safety and independence for the elderly. IoT devices like fall detection sensors, smart home automation, and voice assistants cater to the unique needs of seniors, allowing them to live comfortably and securely in their homes.
10. *Wearable Health Tech for Athletes*: In the sports and fitness domain, HCIoT applications offer wearable health tech for athletes and fitness enthusiasts. Smart athletic wear, such as smart shoes and fitness apparel, can monitor performance metrics and provide real-time feedback during workouts, optimizing training routines and minimizing the risk of injuries.
11. *Smart Healthcare*: In HCIoT-based healthcare applications, IoT devices and wearables monitor patients' health conditions and transmit data to healthcare providers in real time. By designing user-friendly interfaces and integrating patient feedback, HCIoT enhances patient engagement and adherence to treatment plans.
12. *Assistive Technology*: HCIoT plays a crucial role in the development of assistive technologies for individuals with disabilities. IoT-enabled devices, such as smart home systems and wearable assistive devices, empower people with mobility or sensory impairments, allowing them to control their environment and improve their quality of life.
13. *Smart Homes*: HCIoT is evident in smart home applications where users interact with a wide range of connected devices, such as smart thermostats, lighting systems, and home security. By providing intuitive smartphone apps and voice-activated controls, HCIoT simplifies home automation and enhances user convenience.
14. *Personalized Retail Experiences*: Retailers are leveraging HCIoT to create personalized shopping experiences for customers. IoT sensors and

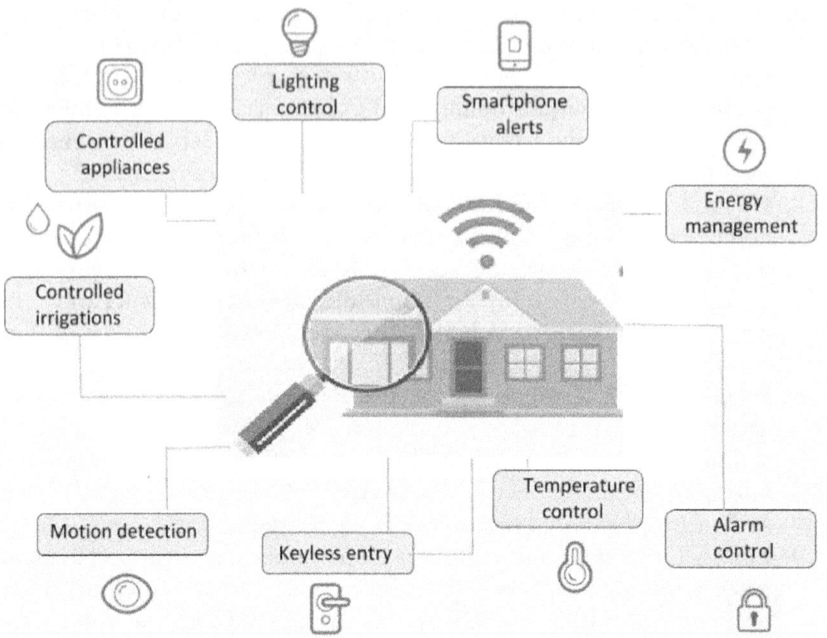

Figure 6.5 Smart homes.

beacons in physical stores can gather data on customer preferences, allowing retailers to offer tailored recommendations and promotions.

15. *Transportation and Mobility Solutions*: In the automotive industry, HCIoT is transforming driving experiences with connected cars and smart transportation systems. User-friendly infotainment interfaces, voice assistants, and context-aware navigation contribute to safer and more enjoyable journeys.

6.13 MULTIMEDIA IoT

Multimedia IoT (M-IoT) devices differ from those created using IoT technology, requiring substantial memory, computational power, and operating as high-power consumers with significant bandwidth needs (Dargie & Poellabauer, 2010). The application of IoT in real-time scenarios spans various domains, including smart cities, smart grids, smart hospitals, industrial IoT, smart homes, and smart agriculture. Key attributes of M-IoT include reliability, timeliness in delivering essential data, enforcement of quality of service (QoS) standards, and the need for an efficient architecture in the transmission network.

User-perceived QoS translates into the quality of experience (QoE), categorized as subjective or objective. The escalating generation of multimedia information poses challenges in storing, processing, sharing, and exchanging data, requiring cloud, edge, and fog computing technologies. The routing protocol for low-power and lossy networks (RPL) is the standardized IoT routing protocol, necessitating further development to address load balancing, fault tolerance, delay, and energy-aware M-IoT deployment scenarios (Jain et al., 2021).

IoT characteristics facilitate communication with multimedia, which is sensitive to delay and bandwidth-intensive. The rapid increase in traffic due to multimedia data in IoT has spurred the development of innovative approaches to meet these demands. M-IoT devices demand ample memory, quick computation, and extensive bandwidth to process data. Communication typically involves multipoint-to-multipoint and multipoint-to-point contexts. Real-world multimedia applications encompass emergency response systems, traffic monitoring, industrial IoT, crime inspection, smart agriculture, smart homes, smart cities, smart hospitals, and surveillance systems (Jain et al., 2021).

6.14 SOCIETAL IMPACT AND SUSTAINABLE DEVELOPMENT

HCIoT and M-IoT have become pivotal technologies with profound societal impacts and implications for sustainable development. As these technologies continue to advance, they contribute significantly to shaping the way we live, work, and interact with the world around us. They include the following:

1. *Enhanced Quality of Life*: HCIoT focuses on creating technology that revolves around human needs and experiences. Smart homes, wearable devices, and health monitoring systems are prime examples. These technologies enhance convenience, safety, and overall well-being, contributing to an improved quality of life for individuals.
2. *Healthcare Revolution*: The integration of HCIoT in healthcare is transformative. Remote patient monitoring, smart medical devices, and personalized healthcare solutions enable timely interventions, reduce hospitalization rates, and enhance the effectiveness of medical treatments.
3. *Urban Development and Smart Cities*: The deployment of IoT in urban planning leads to the development of smart cities. Intelligent transportation systems, energy-efficient infrastructure, and enhanced public services contribute to sustainable urban development, improving the overall living conditions in densely populated areas.
4. *Accessibility and Inclusion*: HCIoT plays a crucial role in promoting accessibility and inclusion. Assistive technologies, such as smart devices for people with disabilities, contribute to a more inclusive society, breaking down barriers and providing equal opportunities for all.

5. *Environmental Monitoring*: IoT technologies, including multimedia IoT, are employed for environmental monitoring and conservation efforts. Sensors and devices gather data on air quality, water resources, and wildlife, enabling informed decision-making and sustainable resource management.

HCIoT and M-IoT can also play a crucial role in sustainable development, including following:

1. *Healthcare*: Multimedia data serves as the conduit for communication, monitoring, and collaboration across diverse facets of daily life at multiple granularity levels across various applications. In the realm of health, this is recognized as personal health media. The integration of personal health media into a single device for comprehensive IoT-based measurements embodies the characteristics of the forthcoming health system (Jain et al., 2021).

   Sensors gather information from input data of doctors, patients, and nurses and AI and machine learning (ML) algorithms are used to examine the gathered data, which is further sent to the devices that determine whether to send data to the cloud. After this, doctors or medical professionals assess the data provided by IoT healthcare solutions to make informed decisions (Vaghela, 2023).
2. *Energy Efficiency*: HCIoT and M-IoT contribute to sustainable development by promoting energy efficiency. Smart buildings, intelligent

Figure 6.6 Precision agriculture.

lighting, and energy management systems optimize resource usage, reducing environmental impact and promoting sustainable practices.

3. *Precision Agriculture*: In agriculture, IoT technologies enable precision farming practices. Smart sensors, drones, and data analytics optimize crop management, reduce resource wastage, and promote sustainable agricultural practices, ensuring food security while minimizing environmental impact. In precision agriculture, IoT/sensor nodes play a crucial role in gathering real-time data, enhancing the practicality of the system. These nodes possess the capability to collect real-time data from crop fields, thereby refining the precision of the agricultural system. The integration of data analytics and machine learning further enhances the functionality of the agricultural system. These technologies find extensive applications in diverse fields. Precision agriculture has led to the development of various applications for farmers, providing timely information about the status of crops (Akhter & Sofi, 2021).
4. *Waste Management*: IoT plays a crucial role in waste management systems. Smart waste bins, tracking systems, and data analytics streamline waste collection processes, reducing litter and promoting recycling efforts, contributing to a cleaner and more sustainable environment.
5. *Supply Chain Optimization*: Both HCIoT and M-IoT contribute to supply chain optimization. Real-time tracking, inventory management,

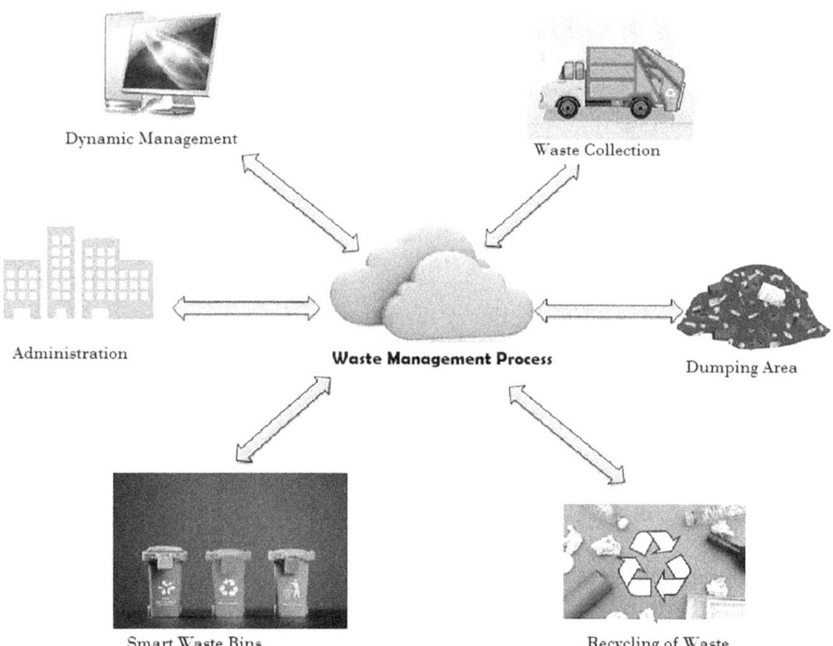

Figure 6.7 Smart waste management.

and data analytics enhance efficiency, reduce waste, and promote sustainable business practices throughout the supply chain. The incorporation and assimilation of IoT have exhibited tangible benefits in social, economic, and environmental aspects. The entire supply chain has undergone a metamorphosis, becoming flexible and dynamic while ensuring high-quality services at a low cost. This chapter provides valuable insights for supply chain policymakers and managers, guiding them through the industrial revolution. The precision required in packaging manufactured products destined for distribution centers necessitates heightened accuracy. In this regard, the extensive deployment of sensory networks, RFIDs, and bar codes within the supply chain proves to be an effective solution, capable of seamless integration with distribution centers for proactive readiness (Khan et al., 2023).

At the supply chain management system's distribution end, careful observation of goods and vehicles, safe delivery of goods to distribution locations, and real-time position updates can significantly reduce losses. Furthermore, the use of smart storage becomes essential for continuously improving product management and space optimization. IoT-enabled integrated solutions for loading, unloading, and warehousing save operating times and promote increased productivity. As a result, changes are made to the store layout to improve foot traffic and encourage customers to make larger purchases.

6. Education and Awareness: These technologies facilitate the dissemination of information and knowledge. M-IoT, in particular, supports interactive and engaging educational content, fostering awareness about sustainable practices and encouraging responsible behavior among individuals and communities.

6.15 CONCLUSION

The dynamic and multidisciplinary discipline of HCI is crucial in determining how people interact with digital technologies. HCI changes as technology progresses to incorporate new methods of interaction, tools, and design ideas. Developing interfaces that are efficient, effective, and user-friendly is the main objective of HCI. To achieve this, the field applies concepts like learnability, consistency, clarity, and user-centered design.

The constant efforts to make interactions more natural, immersive, and inclusive are demonstrated by the developments in HCI, which include natural language processing, gesture detection, virtual and augmented reality, brain-computer interfaces, and more. These developments have improved efficiency and quality of life across a range of industries, including healthcare, agriculture, education, and smart cities. The rise of human-centric Internet of Things underscores how essential it is to take human requirements, preferences, and emotions into consideration while designing and implementing IoT systems. HCIoT wants to maximize the potential of IoT technology by prioritizing user-centered design, context awareness, simplicity, and inclusivity. This leads to the creation of meaningful and personalized interactions between people and connected objects.

Even more smooth and organic interactions between people and technology are expected to define the future of HCI. HCI is at the forefront of developing interfaces that not only satisfy functional requirements but also improve the completely human experience in the increasingly digital and interconnected world, as designers and researchers continue to investigate novel solutions and tackle obstacles. The interdependent relationship between multimedia IoT and human-centric IoT has the potential to completely transform sustainable development. In order to create a future in which technology acts as a catalyst for positive societal effect and sustainable progress, we must confront obstacles, think about the ethical implications, and encourage creativity as we traverse this profoundly changing world.

REFERENCES

Akhter, R., & Sofi, S. A. (2021). Precision agriculture using IoT data analytics and machine learning. *Journal of King Saud University—Computer and Information Sciences*, 5603–5618.

Berzowska, J. (2005). Memory rich clothing: Second skins that communicate physical memory. In *Proceedings of the 5th Conference on Creativity & Cognition* (pp. 32–40). ACM Press.

Dargie, W., & Poellabauer, C. (2010). *Fundamentals of Wireless Sensor Networks: Theory and Practice*. John Wiley & Sons.

Jain, A., Yadav, K., Alharbi, Y., Alferaidi, A., Alkwai, L. M., Ahmed, N. M., & Hamad, S. A. (2021, October 25). Current and potential applications of IoT in multimedia communication system. *Research Square*. https://doi.org/10.21203/rs.3.rs-946443/v1

Khan, Y., Su'ud, B. M., Alam, M. M., Ahmad, A. F., Ahmad (Ayassrah), A., & Khan, N. (2023). *Application of Internet of Things (IoT) in Sustainable Supply Chain Management*. www.mdpi.com/2071-1050/15/1/694

Vaghela, P. (2023, May 11). *Iot In Healthcare: Applications, Benefits, Challenges & Future*. www.bigscal.com/blogs/healthcare-industry/iot-in-healthcare-applications-benefits-challenges-and-future/.

Chapter 7

XAI-Driven Approaches for Ensuring Security and Data Protection in IoT

Amita Sharma, Navneet Sharma, and Anubha Jain

7.1 INTRODUCTION

The Internet of Things (IoT) has revolutionized the way we interact with the digital world, seamlessly connecting an extensive array of devices to enhance our daily lives. This vast network of interconnected smart devices has enabled unprecedented convenience, efficiency, and data-driven insights across various industries. In 2020, Cisco estimated a staggering 50 billion connected IoT devices, surpassing the world population. These devices generate vast amounts of data, following three-layer architecture: physical, network, and application. Smart home systems, like fridges placing orders automatically, and smart hospitals, monitoring patients in emergencies, exemplify IoT applications (Čolaković & Hadžialić, 2018). However, with this transformational growth comes a multitude of security challenges that demand immediate attention. The implementing IoT poses numerous challenges, such as standardization, data management, trust, security, and privacy across various IoT applications.

Additionally IoT systems are vulnerable to an array of security issues, making them enticing targets for cyberattacks. Device vulnerabilities, data breaches, and privacy infringements pose significant threats, potentially leading to data compromise, system failures, and even physical harm. The application of artificial intelligence (AI) has emerged as a key enabler in fortifying IoT systems against security threats (Mohanta et al., 2020). AI's ability to analyze vast amounts of data, detect patterns, and make autonomous decisions has proven invaluable in various domains. Within the realm of IoT, AI offers promising solutions for intrusion detection, anomaly identification, and predictive maintenance, among others.

Common IoT attacks include Frantic Locker (FLocker) targeting smart TVs, the infamous Mirai botnet, and those exploiting vulnerabilities in smart bulbs. Consequences range from device control to network disruption (Abed & Anupam, 2023). AI plays a crucial role in avoiding IoT cyberattacks by using anomaly detection to identify abnormal behavior, behavior analysis to detect deviations, and predictive analytics to anticipate potential threats. AI can also strengthen access control mechanisms, detect and

DOI: 10.1201/9781003406723-7

neutralize malware, and provide real-time response to mitigate the impact of attacks. Its ability to analyze vast amounts of data and learn from patterns is vital in safeguarding IoT systems against evolving cyber threats (Sarker et al., 2022). Leveraging the proficiency of AI, particularly machine and deep learning solutions, holds the key to providing a dynamically enhanced and continuously updated security system for the future generation of IoT.

AI-based solutions for IoT security can be classified into five categories. First, anomaly detection, where AI analyzes IoT data to identify abnormal behavior and potential threats. Second, behavioral analysis, where AI learns the normal behavior of devices and users to detect deviations. Third, dynamic access control, where AI evaluates access requests in real time based on contextual information. Fourth, threat intelligence, where AI continuously monitors and analyzes global threat data. Fifth, automated response, where AI enables swift actions to contain security incidents and minimize their impact. AI empowers IoT security by efficiently processing vast amounts of data, enabling real-time decision-making, and proactively adapting to emerging threats. Leveraging the proficiency of AI, particularly machine and deep learning solutions, holds the key to providing a dynamically enhanced and continuously updated security system for the future generation of IoT.

Traditional AI has been harnessed to bolster IoT security, but its black box nature leaves a veil of opacity over the decision-making process, hindering our ability to fully comprehend and trust its actions. Often, AI algorithms operate as inscrutable black boxes, making it difficult to decipher the rationale behind their decisions (Lipton, 2018). This opacity raises concerns about trust, accountability, and potential biases within AI-driven security measures.

In the rapidly evolving landscape of advanced technology, the advent of explainable artificial intelligence (XAI) has stirred considerable interest across industry and academia alike. With relentless efforts, XAI has achieved noteworthy success, producing models that inspire confidence through their transparent decision-making processes (Mohseni et al., 2021). XAI plays a crucial role in enhancing IoT security solutions by providing transparency and interpretability to AI-driven models and decisions. In widely used XAI services for IoT applications, such as security enhancement, Internet of Medical Things (IoMT), Industrial IoT (IIoT), and Internet of City Things (IoCT), the ability to understand the reasoning behind AI-based actions becomes essential (Jagatheesaperumal et al., 2022).

In IoT security, XAI empowers security teams to comprehend how AI models identify potential threats and anomalies. This understanding is critical in validating the reliability of security measures and building trust in the decision-making process. For example, in IoMT, where medical devices communicate with each other and collect sensitive patient data, XAI can explain the factors that contribute to an alert or diagnosis, ensuring medical professionals can trust and verify the AI-driven insights (Sinha et al., 2023).

In IIoT and IoCT applications, where critical infrastructure and urban systems are involved, XAI's transparency allows for real-time monitoring of

AI models' actions and enhances the ability to detect cyber threats or anomalies promptly (Zolanvari et al., 2021). This enables a proactive response to potential security breaches, ensuring the robustness of industrial processes and city services. XAI encompasses various models such as local interpretable model-agnostic explanations (LIME), SHapley Additive exPlanations (SHAP), and rule-based models (Arrieta et al., 2020). These models shed light on how AI decisions are reached in IoT security, offering transparency. By incorporating XAI in these IoT applications, security becomes more explainable, actionable, and effective, ultimately bolstering the overall trustworthiness and reliability of AI-based IoT security solutions.

Throughout this chapter, we will explore the significant implications of XAI in mitigating IoT security risks. We will use the case study examples from smart hospitals, smart cities, and smart homes to show how AI's interpretability helps security professionals detect and respond to threats effectively. We will also discuss the challenges of implementing XAI in IoT security, like technical complexities and ethical concerns. In this chapter's conclusion, we'll give you an idea of XAI's potential applications in IoT security in the future. As the technology advances, it has the potential to reshape the landscape of IoT security and usher in a new era of trust, transparency, and protection in our linked world.

7.2 IoT SYSTEMS AND SECURITY ISSUES

As of 2023, IoT has experienced exponential growth, revolutionizing industries and permeating various aspects of our daily lives. It has evolved into a vast ecosystem of interconnected devices and systems, generating an unparalleled amount of data that fuels automation, analytics, and innovation across diverse domains. From healthcare to manufacturing, smart homes to smart cities, the applications and scope of IoT have expanded significantly, promising enhanced efficiency, convenience, and sustainability. The proliferation of IoT devices and applications has been remarkable in 2023, fueled by advancements in connectivity, sensor technology, cloud computing, and artificial intelligence. The number of connected devices has surged beyond expectations, with billions of devices seamlessly communicating with each other and interacting with users. This growth has been particularly prominent in industries like healthcare, agriculture, transportation, and smart infrastructure, where IoT's transformative potential is being harnessed to optimize operations, improve decision-making, and drive innovation (Rejeb et al., 2023; Fotia et al., 2023).

7.2.1 Applications and Scope in Different Domains

In healthcare, IoT-enabled medical devices and wearables have revolutionized patient monitoring and telemedicine, facilitating remote health management and improving patient outcomes. Agriculture has witnessed the deployment of IoT sensors for precision farming, enabling farmers to optimize water usage, monitor soil conditions, and boost crop yields (McCaig et al., 2023). In transportation, IoT technologies have paved the way for connected vehicles and smart transportation systems, promoting safer and more efficient mobility. Smart homes have become commonplace, with IoT devices enhancing energy efficiency, security, and home automation. In urban planning, the concept of smart cities has gained traction, leveraging IoT to optimize resources, enhance citizen services, and create sustainable urban environments (Fizza et al., 2022).

7.2.2 Security Challenges in IoT

Amidst the incredible potential of IoT, security remains a major challenge that hinders its widespread adoption and further growth. The unique characteristics of IoT systems, such as heterogeneous devices, constrained resources, and diverse communication protocols, create vulnerabilities that malicious actors can exploit. IoT security issues can be categorized into the following five categories:

- Device Vulnerabilities: IoT devices often lack robust security features, making them susceptible to hacking and unauthorized access. Weak passwords, outdated firmware, and unpatched vulnerabilities are common issues that leave devices open to exploitation. Weaknesses in IoT device security can be addressed through better authentication, secure boot processes, and regular software updates to patch known vulnerabilities (Feng et al., 2022).
- Data Privacy and Encryption: The massive volume of data generated by IoT devices presents data privacy concerns. Inadequate encryption during data transmission and storage can lead to unauthorized access and data breaches, compromising user privacy. Implementing strong encryption algorithms for data transmission and storage, along with data anonymization techniques, can safeguard user privacy and protect sensitive information (Rathore et al., 2022).
- Network Security: IoT networks are susceptible to attacks such as distributed denial of service (DDoS) and man-in-the-middle (MITM) attacks, which can disrupt communication and compromise data integrity. Deploying firewalls, intrusion detection systems, and network segmentation can help protect IoT networks from unauthorized access and mitigate DDoS and MITM attacks (Saba et al., 2022).
- Authentication and Authorization: Inadequate authentication mechanisms can lead to unauthorized access to IoT devices and systems. Without proper authorization protocols, malicious actors can gain control over critical infrastructure or sensitive information. Implementing multi-factor authentication and role-based access control

ensures only authorized users have access to IoT devices and systems (Istiaque et al., 2021).

- Physical Security and Safety: IoT systems that control physical devices, such as smart homes and industrial IoT applications, are at risk of physical tampering or attacks, which can have real-world consequences and pose safety risks. Ensuring physical security measures, such as tamper-resistant designs and access controls, can safeguard IoT systems from physical attacks and ensure user safety (Yang et al., 2022).

Figure 7.1 shows the key issues associated with IoT security.

Indeed, the low power and limited computational capabilities of IoT devices pose significant challenges when it comes to cybersecurity. These constraints make IoT devices attractive targets for cyberattackers as they are often easier to compromise compared to traditional computing systems. Some of the common cyberattacks that exploit these vulnerabilities include the following:

- Botnet Attacks: IoT devices can be harnessed to form botnets, which are large networks of compromised devices controlled by malicious actors. These botnets can be used to launch DDoS attacks,

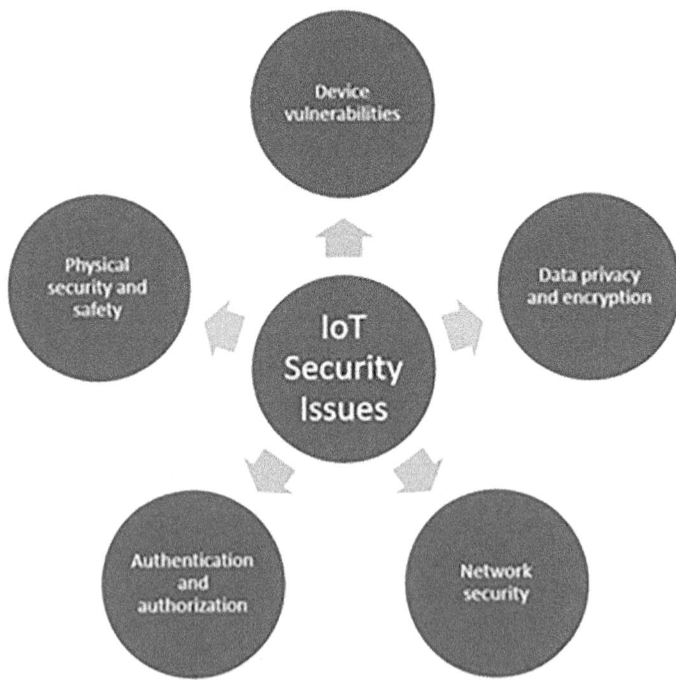

Figure 7.1 IoT security issues.

overwhelming servers with a massive volume of traffic and causing disruptions to online services (Alazzam et al., 2019).

- Ransomware: IoT devices may fall victim to ransomware attacks, where malicious software encrypts the device's data, rendering it inaccessible to the owner. Attackers then demand a ransom to provide the decryption key, extorting money from the device's owner (Humayun et al., 2021).
- Man-in-the-Middle Attacks: IoT devices communicating over unsecured networks can be susceptible to MITM attacks, where an attacker intercepts and potentially alters the data transmitted between devices, compromising the integrity and confidentiality of the information (Mallik, 2019).
- Eavesdropping and Data Interception: Weak encryption or lack of encryption on IoT devices may enable attackers to eavesdrop on communication and intercept sensitive data, such as personal information or authentication credentials.
- Physical Tampering: Physical access to IoT devices can allow attackers to tamper with the device's hardware or software, potentially compromising its functionality or extracting sensitive data (Rejeb et al., 2019).

Addressing these cybersecurity challenges requires a multi-faceted approach, considering the unique limitations of IoT devices. Manufacturers must prioritize security in the design phase, implementing robust authentication mechanisms, encryption protocols, and secure boot processes. Regular software updates and security patches should be provided to address known vulnerabilities and improve device resilience. Thus, the growth and potential of IoT in 2023 are awe-inspiring, transforming diverse domains with unprecedented connectivity and automation. However, IoT security challenges remain a critical impediment to its widespread adoption. Categorizing and addressing security issues through robust authentication, encryption, network protection, and physical security measures are essential steps to unlock the full potential of IoT while ensuring its safe and secure integration into our daily lives.

Numerous techniques are utilized to tackle IoT security challenges. These encompass encryption algorithms for secure data transmission, authentication mechanisms for device verification, intrusion detection systems to spot malicious activities, access controls for restricting unauthorized access, and anomaly detection to identify abnormal behavior. Additionally, secure communication protocols, regular updates, and robust security policies reinforce IoT system security. The incorporation of AI further empowers IoT security through intelligent analytics and machine learning algorithms, detecting anomalies, identifying threats, enhancing authentication, and automating security operations to fortify the overall resilience of IoT systems. The next section discusses AI's potential in handling IoT security issues.

7.3 AI APPLICATIONS IN IoT

AI is the simulation of human intelligence in machines that can learn, reason, and make decisions. It is an enthralling amalgamation of cutting-edge technologies, encompassing machine learning, neural networks, and natural language processing. This captivating discipline endows machines with cognitive abilities, enabling them to analyze data, discern patterns, and make informed decisions, unlocking transformative potential across diverse industries. Its potential lies in transforming various industries, including IoT, by augmenting the capabilities of devices and systems, enabling them to process vast amounts of data, identify patterns, and adapt to dynamic environments. AI's ability to analyze and derive insights from data holds significant promise for enhancing IoT applications and addressing various challenges.

7.3.1 AI Enhances IoT in Different Areas

The following are a few examples of how AI has aided the growth of IoT:

- Predictive Maintenance: AI-driven predictive maintenance in IoT enables the identification of potential equipment failures before they occur. By analyzing sensor data, AI can predict maintenance needs and schedule timely repairs, reducing downtime and optimizing resource utilization (Rojek et al., 2023).
- Energy Management: AI can optimize energy consumption in IoT systems by analyzing data from smart meters and devices. It can identify usage patterns and implement energy-saving strategies, leading to increased energy efficiency and cost savings (Bedi et al., 2022).
- Healthcare and Remote Monitoring: AI-powered IoT devices in healthcare enable remote patient monitoring, real-time diagnostics, and personalized treatment plans. AI algorithms can analyze health data to detect abnormalities, allowing for early intervention and improved patient outcomes (Jamil et al., 2020).
- Smart Homes: AI enhances smart home systems by understanding user preferences and habits. Through machine learning, IoT devices can anticipate user needs, such as adjusting temperature, lighting, and entertainment, providing a seamless and personalized experience (Azumah et al., 2021).
- Traffic Management: AI in IoT-based traffic management systems can optimize traffic flow, reduce congestion, and enhance road safety. AI algorithms analyze data from sensors and cameras to make real-time traffic predictions and optimize signal timings (Chavhan et al., 2022).
- Agriculture: AI-powered IoT sensors and drones in agriculture provide real-time data on soil conditions, weather patterns, and crop health. AI analysis enables precision agriculture practices, such as targeted irrigation and fertilization, leading to improved crop yields (Rehman et al., 2022).

7.3.2 Limitations of AI in Handling IoT Security

Despite being the most promising technology, AI still has many limitations. In the following, we will discuss how AI struggles to address IoT security challenges (Singh et al., 2020).

- Resource Constraints: Many IoT devices have limited computational power and memory, making it challenging to deploy complex AI algorithms. Simplified AI models may compromise the accuracy and effectiveness of security measures.
- Latency: Real-time AI-based security analysis in IoT can introduce latency, affecting critical applications like healthcare and autonomous vehicles. Reducing latency while maintaining accurate results remains a challenge.
- Data Privacy and Security: AI requires extensive data to train models effectively. Ensuring data privacy and security in IoT, especially when dealing with sensitive information, becomes crucial to prevent unauthorized access.
- Adversarial Attacks: AI models are vulnerable to adversarial attacks, where malicious inputs can deceive the models and compromise security decisions. In IoT systems, adversarial attacks can lead to unauthorized access and control over devices.
- Explainability: Explainable AI is vital in security-critical IoT applications. The lack of interpretability in AI models can hinder the understanding of security decisions, making it difficult to trust the outcomes.

AI's integration in IoT has opened new avenues for innovation and efficiency across various sectors. From predictive maintenance to healthcare and smart homes, AI empowers IoT applications with intelligent data analysis and decision-making capabilities. However, several limitations, such as resource constraints, data privacy, and adversarial attacks, need to be addressed for a secure and robust IoT ecosystem. By overcoming these challenges, AI-driven IoT security can be elevated to provide trustworthy and resilient protection for our interconnected world.

7.4 EXPLAINABLE ARTIFICIAL INTELLIGENCE

Explainable artificial intelligence is an innovative branch of AI that aims to provide transparency and interpretability to AI-driven models and decisions. Unlike traditional black box AI, XAI empowers users to understand how AI systems arrive at specific outcomes, uncovering the underlying reasoning and factors that influence their decisions. By providing human-readable explanations, XAI not only enhances the trustworthiness of AI applications but also allows for better insight into potential biases, vulnerabilities, and risks. With XAI, users can make informed decisions, identify system weaknesses, and take proactive measures, fostering responsible and ethical use of AI in various domains.

The quest for explainability arises from the growing adoption of AI in critical applications, where understanding the reasons behind AI-generated decisions is crucial for building trust, ensuring fairness, and meeting regulatory requirements. Various XAI techniques have been developed, each catering to specific applications and scenarios. In this discourse, we explore some of the prominent XAI techniques and their diverse applications.

1. *Local Interpretable Model-Agnostic Explanations (LIME)*: LIME is a popular XAI technique that provides explanations for individual predictions made by complex machine learning models. It approximates the behavior of the black box model locally around the prediction of interest by generating a simplified interpretable model (Arrieta et al., 2020). The simplified model, typically a linear model, is easier to understand and reveals the features most influential in driving the model's decision. LIME has found applications in various domains, such as image classification, natural language processing, and medical diagnosis, where understanding specific model predictions is crucial for users.
2. *SHapley Additive exPlanations (SHAP)*: SHAP is rooted in cooperative game theory and provides a unified framework for interpreting the output of machine learning models. It assigns contributions to each feature in the input data based on its importance in the model's prediction. By capturing feature interactions and dependencies, SHAP offers a comprehensive understanding of the model's behavior and aids in identifying which features influence predictions positively or negatively (Jagatheesaperumal et al., 2022). This technique finds applications in finance, healthcare, and recommendation systems, where accurate feature-level explanations are essential for decision-making.
3. *Rule-Based Explanations*: Rule-based explanations involve generating human-readable rules that mimic the decision-making process of a black box model. These rules provide clear, interpretable explanations for individual predictions (Macha et al., 2022). Rule-based explanations are often used in decision trees and rule-based classifiers and are valuable in domains where users require simple, transparent explanations, such as credit risk assessment and loan approval systems.
4. *Counterfactual Explanations*: Counterfactual explanations generate alternative input instances that would have led to a different model prediction. By showing how slight changes in the input data impact the output, counterfactual explanations offer insights into the model's sensitivity to different features (Nwakanma et al., 2023). This technique is particularly useful in applications like healthcare, where understanding how small changes in patient data affect treatment recommendations is critical.

5. *Certifiable Explanations*: Certifiable explanations focus on providing guarantees about the reliability and robustness of model explanations. These techniques aim to ensure that the generated explanations are not only accurate but also sufficiently robust to variations in the input data (Landgrebe, 2022). Certifiable explanations find applications in safety-critical domains, such as autonomous vehicles and medical diagnosis, where model accountability and reliability are paramount.

XAI encompasses a rich array of techniques designed to provide interpretable and human-understandable explanations for AI-driven decisions. Each XAI technique caters to specific applications and scenarios, empowering users to gain insights into the inner workings of complex AI models (Sharma, 2023). By fostering transparency and trust in AI systems, XAI plays a pivotal role in ensuring the responsible and ethical deployment of AI across various domains, unlocking the full potential of AI as a transformative force for positive change.

7.5 SECURITY ISSUES ADDRESSED BY XAI IN IoT

As the IoT continues to proliferate, the security of IoT devices and systems becomes a paramount concern. Device vulnerabilities, data privacy and encryption, network security, authentication and authorization, and physical security and safety are key issues that demand effective solutions. While AI has been utilized to address these challenges, XAI offers more transparent and interpretable insights, providing superior solutions for enhancing IoT security.

- Device Vulnerabilities: AI has been employed to detect potential device vulnerabilities, but it may not fully understand the intricacies of IoT devices and the specific security features they lack. In contrast, XAI techniques like counterfactual explanations offer clear insights into device vulnerabilities (Nwakanma et al., 2023). For instance, consider a smart home system with multiple IoT devices. XAI can analyze the decisions made by the AI model and highlight the weak points in individual devices' security, such as outdated firmware or weak passwords. This enables security professionals to implement targeted measures and prioritize security patches more effectively, bolstering the overall resilience of IoT devices.
- Data Privacy and Encryption: AI-based encryption may be vulnerable to attacks or may not effectively handle data anonymization, leading to unauthorized access to sensitive information. On the other hand, XAI, particularly SHAP, excels in enhancing data privacy and encryption. For example, in a healthcare IoT application, SHAP can identify critical patient data points that need stronger encryption and anonymization. This ensures that confidential medical information remains protected from unauthorized access and maintains patients' privacy.

- Network Security: AI has been utilized to detect and respond to network threats, but it may not promptly identify sophisticated attacks like DDoS and MITM. XAI techniques, like LIME, offer insights into network security issues. In an industrial IoT scenario, LIME can reveal how a malicious actor is exploiting the network architecture to launch a DDoS attack on critical infrastructure (Zolanvari et al., 2021). This enables security teams to deploy targeted countermeasures and protect against potential network disruptions.
- Authentication and Authorization: AI-driven authentication mechanisms may not be robust enough to prevent unauthorized access to IoT devices and systems. XAI, with the assistance of SHAP, can enhance authentication and authorization measures. For instance, in a smart city application, SHAP can identify potential weak points in user authentication, such as insufficient user verification methods. By providing such insights, XAI empowers administrators to implement multi-factor authentication and role-based access controls, bolstering the security of the smart city infrastructure (Srivastava et al., 2022).
- Physical Security and Safety: AI may lack an understanding of the physical security requirements for IoT devices, potentially leaving them vulnerable to tampering or physical attacks. XAI techniques like rule-based methods can offer valuable insights into the physical security needs of IoT systems (Rawal et al., 2021). In an autonomous vehicle context, they can reveal potential vulnerabilities in the vehicle's physical design, such as exposed ports or unsecured access panels. This enables engineers to implement tamper-resistant designs and access controls, ensuring the safety and security of autonomous vehicles.

While AI has been beneficial in addressing IoT security challenges, XAI provides a more profound and transparent understanding of the decision-making process of AI models. XAI techniques like LIME and SHAP offer interpretable insights into device vulnerabilities, data privacy and encryption, network security, authentication and authorization, and physical security and safety. Table 7.1 illustrates the key issues with possible XAI based solutions. By leveraging XAI, security professionals can make informed decisions, proactively address vulnerabilities, and implement targeted security measures to fortify IoT devices and systems against potential cyber threats. The future of IoT security lies in the responsible use of AI and XAI, paving the way for a safer and more secure interconnected world.

7.6 CASE STUDIES

In this section, four particular case studies are discussed along with an explanation of how XAI would enhance these systems' overall performance.

Table 7.1 Key Security Issues Addressed by XAI in IoT

Key Issue	*Description*	*How AI Lacks*	*How XAI Solves*	*XAI Algorithm*
Device Vulnerabilities	Lack of robust security features in IoT devices	AI may not understand device vulnerabilities and may not prioritize security measures effectively.	XAI provides transparent insights into device vulnerabilities, allowing for proactive security measures.	Counterfactual explanations
Data Privacy and Encryption	Inadequate data protection during transmission and storage	AI may not effectively handle encryption and data anonymization, leading to potential unauthorized access and breaches.	XAI employs strong encryption algorithms and data anonymization techniques for enhanced data privacy and security.	SHAP
Network Security	Vulnerabilities to DDoS and MITM attacks	AI may not have the capability to detect sophisticated network attacks promptly.	XAI uses intrusion detection and network segmentation to swiftly identify and mitigate network security threats.	LIME
Authentication and Authorization	Weak authentication mechanisms and unauthorized access	AI may lack the ability to enforce robust authentication protocols effectively.	XAI implements multi-factor authentication and role-based access control for enhanced user authorization.	SHAP
Physical Security and Safety	Risk of physical tampering and attacks	AI may not comprehend physical security needs and may not provide sufficient safeguards against tampering.	XAI ensures physical security measures, such as tamper-resistant designs and access controls, to protect against physical attacks.	Rule-based methods

7.6.1 Project in Italy Smart Agriculture

In an agricultural setting in Italy, IoT devices were deployed to monitor soil conditions, control fertilizer application, manage irrigation, and monitor plant growth (Vanino et al., 2018). Traditional AI models were utilized to optimize these processes, but their decisions were often regarded as black box solutions, making it challenging for farmers to comprehend the reasoning behind recommendations.

By integrating XAI into the IoT system, farmers gained valuable insights into the decisions made by the AI models. Techniques like LIME provided clear explanations for specific plant growth predictions and recommendations on fertilizer usage. Farmers could understand the factors influencing the model's decisions, such as soil moisture, nutrient levels, and weather conditions. With this transparency, farmers will be inclined to trust and act on the AI recommendations, leading to optimized crop yield and resource utilization.

7.6.2 Smart Marina: Monitoring Mooring Berths, Sea Level, and Weather Conditions

In a smart marina project, IoT sensors were installed to monitor mooring berths, sea level, and weather conditions (Maglić et al., 2021). The data collected from these sensors was used to optimize berth allocation, predict adverse weather conditions, and enhance safety for vessels and their occupants.

To make the marina's operations more transparent, XAI can be introduced. The XAI techniques, such as a rule-based approach, can explain the decision-making process behind berth allocations based on factors like vessel size, arrival time, and weather conditions. Boat owners and harbor authorities could be able to comprehend the reasoning behind berth assignments, leading to improved customer satisfaction and better planning during adverse weather conditions.

7.6.3 Pfizer Drug Development

Pharmaceutical companies like Pfizer utilize IoT in drug development, where AI-driven models analyze vast amounts of data to identify potential drug candidates (Alagarsamy et al., 2019). However, the lack of interpretability in traditional AI models poses challenges in understanding drug efficacy and safety (Muhsen et al., 2021).

By adopting XAI with feature importance techniques, Pfizer can gain insights into how specific features influence drug predictions. For example, SHAP and counterfactual explanations can explain the contributions of different chemical properties to drug efficacy. This transparency enables pharmaceutical researchers to validate model decisions, identify potential biases or errors, and refine drug development processes with greater confidence.

7.6.4 Turning a Stadium into a Smart Building—Honeywell

In the context of a smart stadium project, IoT devices were used to monitor various aspects such as energy consumption, occupancy levels, and environmental conditions. AI algorithms were employed to optimize energy usage and enhance stadium operations (Gupta et al., 2021; Gruszka et al., 2017). However, the lack of interpretability in traditional AI models hindered effective decision-making and limited the ability to identify areas for improvement.

By integrating XAI into the smart stadium system, XAI techniques like LIME could explain AI-driven decisions, such as energy optimization strategies based on occupancy patterns and weather forecasts. This transparency empowers stadium managers to comprehend the reasoning behind energy-saving measures and make data-driven decisions for optimal stadium operations, ultimately leading to increased efficiency and cost savings.

Four cases from various disciplines were discussed. We are able to clearly observe that XAI can improve decision-making and the justification for that decision in each case study. However, since real-world problems involve several elements, building a customized solution necessitates the use of a multi-intelligent agent. If XAI is introduced to such systems, they will become more resilient. For improved administration in this situation, explainable multi-agent systems are required.

7.7 EXPLAINABLE MULTI-AGENT SYSTEMS FOR IoT

Explainable multi-agent systems (EMAS) are a specialized approach that brings together multi-agent systems and XAI to enhance transparency and interpretability in complex IoT environments. As IoT applications become more prevalent and mission critical, the need to understand the decision-making process of AI-driven agents becomes paramount. EMAS provides a powerful solution to this challenge, offering clear and human-understandable explanations for the actions and decisions of multiple intelligent entities working collaboratively in an IoT ecosystem (Alzetta et al., 2020).

EMAS for IoT consists of several essential components that contributes to its effectiveness and transparency (Figure 7.3).

- *Agent Architecture*: At the core of EMAS for IoT lies the agent architecture, which encompasses individual agents representing IoT devices or entities. These agents possess cognitive abilities to process information, reason, and make decisions autonomously. A distributed architecture allows agents to collaborate, share information, and collectively achieve goals. Each agent's behavior is governed by rules, strategies, or learning algorithms that can be scrutinized for transparency.

- *Communication Module*
  - *Communication Infrastructure*: Effective communication among agents is vital for collaborative decision-making in EMAS for IoT. A robust communication infrastructure facilitates the exchange of information, negotiation, and coordination. Communication protocols and message formats support agents in sharing explanations for their actions, thereby enabling a comprehensive view of the system's behavior (Zouai et al., 2017).
  - *User Interface and Interaction Module*: To bridge the gap between the technical complexities of EMAS for IoT and human users, a user interface and interaction module is essential. This component presents explanations, insights, and aggregated data to users in an intuitive manner. It enables users to query the system, receive explanations for actions, and make informed. Visualization tools and interactive interfaces facilitate a user-friendly experience.
- *Knowledge and Inference Module*
  - *Contextual Knowledge Base*: EMAS for IoT leverages a contextual knowledge base that stores relevant information about the environment, device capabilities, user preferences, and historical interactions. This knowledge base enriches agents' decision-making processes and assists in generating meaningful explanations (Suwardi & Surendro, 2014). Machine learning techniques can be employed to continuously update and refine the knowledge base based on new data.
  - *Explainability Module*: The explainability module is a pivotal component that enhances the transparency of agent actions and

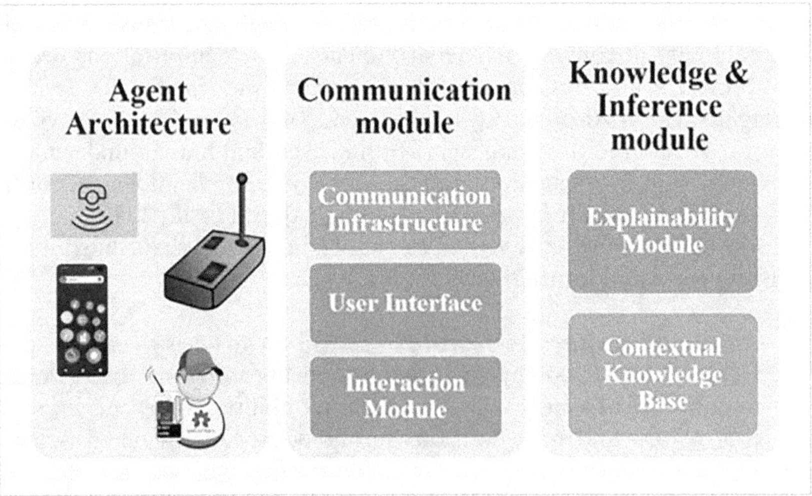

Figure 7.2 Explainable multi-agent systems for IoT.

decisions. It employs a variety of techniques, such as rule-based explanations, model interpretability methods, and natural language generation, to produce human-understandable explanations for agent behavior. This enables stakeholders to comprehend the rationale behind IoT device interactions and system-level outcomes.

Briefly, EMAS for IoT integrates various components to create a transparent and comprehensible framework for managing the intricacies of IoT environments. By combining agent intelligence, explainability techniques, communication capabilities, contextual knowledge, and user-friendly interfaces, EMAS empowers users to better understand and trust the behavior of interconnected IoT devices. As the IoT continues to evolve, the development of robust EMAS will play a pivotal role in ensuring the responsible and accountable deployment of IoT technologies.

7.8 ETHICAL CONSIDERATIONS AND FUTURE PROSPECTS

The rise of IoT has led to significant advancements in various domains, but it also brings along serious security concerns. As IoT devices become more integrated into our daily lives, safeguarding sensitive data and ensuring system reliability is of paramount importance. To address these challenges, XAI and XAI-based multi-agent systems have emerged as potential solutions. This section explores the ethical considerations surrounding the implementation of XAI and XAI-based multi-agent systems in IoT security and discusses their promising future prospects.

7.8.1 Ethical Considerations

There are certain ethical considerations in XAI adaptation. In the following, we will discuss them in detail.

- Transparency and Accountability: XAI aims to make AI models interpretable and explainable to users, enabling them to understand the reasoning behind AI-driven decisions. In the context of IoT security, transparency is vital to assure users that their data is handled responsibly. However, striking the right balance between transparency and security is essential, as disclosing certain details could expose vulnerabilities to malicious actors (Jagatheesaperumal et al., 2022).
- Bias and Fairness: AI models often inherent biases from training data, which could lead to discriminatory outcomes. In the context of multi-agent systems in IoT security, biased decisions could disproportionately impact certain individuals or groups, leading to privacy breaches

or unfair treatment. Ensuring fairness in AI models is crucial for building trust and maintaining societal harmony (Albahri et al., 2023).

- Privacy and Data Protection: As XAI facilitates better understanding of AI models, it becomes crucial to safeguard the privacy of individuals whose data is being processed. IoT devices collect vast amounts of personal data, and ensuring its protection becomes a significant ethical concern. Striking a balance between collecting sufficient data for AI models to make accurate decisions while respecting user privacy is a challenge that needs to be addressed.
- Informed Consent: Implementing XAI-based multi-agent systems in IoT security requires users to be informed about how their data is used, processed, and shared (Amann et al., 2020). Obtaining informed consent becomes crucial in ensuring users' autonomy and agency over their data. However, explaining complex AI systems to non-experts poses communication challenges, demanding user-friendly explanations.

7.8.2 Future Prospects

The following are key areas for future exploration in XAI with IoT:

- Enhanced User Trust and Acceptance: By providing explanations for AI-driven decisions, XAI enhances user trust in AI systems and IoT devices. Understanding why a system made a particular choice fosters user confidence, leading to greater acceptance and adoption of AI-based security measures.
- Improved System Robustness: XAI helps identify vulnerabilities and biases in AI models, leading to more robust systems. By highlighting potential weaknesses, XAI allows developers to fine-tune models, reducing the risk of exploitation by malicious actors.
- Regulatory Compliance: As data privacy and AI ethics gain increasing attention, XAI's adoption can help organizations comply with stringent regulations and standards related to data protection and ethical AI usage. This compliance will be crucial for businesses operating in the IoT security domain, as non-compliance could lead to severe legal and reputational consequences.
- Human-AI Collaboration: XAI facilitates human-AI collaboration, where experts and non-experts can work together to improve the performance and security of AI systems. This interdisciplinary approach allows domain experts to provide insights, ensuring AI systems remain aligned with real-world requirements.
- Ethical AI Deployment: By addressing the ethical considerations inherent in IoT security, XAI fosters the development and deployment of AI technologies in an ethical and responsible manner. This, in turn, can mitigate potential harms associated with AI-powered IoT systems.

The integration of XAI and XAI-based multi-agent systems in IoT security presents both ethical challenges and promising future prospects. Transparent and accountable AI systems can build user trust, improve system robustness, and ensure regulatory compliance. However, ethical considerations surrounding bias, fairness, privacy, and informed consent must be carefully addressed to maximize the potential benefits while minimizing risks. As the field of AI and IoT security continues to evolve, a responsible and ethical approach to deploying XAI is essential for creating a safer and more secure connected world.

7.9 CONCLUSION

This chapter discusses the dynamic fusion of IoT, AI, and XAI, shedding light on the paramount security challenges within IoT systems. It underscores AI's pivotal role in IoT applications, underscoring XAI's vital contribution to transparency. The exploration of security concerns, showcased through real-world instances like smart irrigation and drug design, reinforces the value of XAI. Furthermore, the emergence of explainable multi-agent systems offers a tangible solution for IoT complexities.

REFERENCES

Abed, A. K., & Anupam, A. (2023). Review of security issues in internet of things and artificial intelligence-driven solutions. *Security and Privacy*, 6(3), e285.

Alagarsamy, S., Kandasamy, R., Subbiah, L., & Palanisamy, S. (2019). Applications of internet of things in pharmaceutical industry. SSRN 3441099. https://dx.doi.org/10.2139/ssrn.3441099

Alazzam, H., Alsmady, A., & Shorman, A. A. (2019, December). Supervised detection of IoT botnet attacks. In *Proceedings of the Second International Conference on Data Science, E-Learning and Information Systems,* pp. 1–6.

Alzetta, F., Giorgini, P., Najjar, A., Schumacher, M. I., & Calvaresi, D. (2020, May). In-time explainability in multi-agent systems: Challenges, opportunities, and roadmap. In *International Workshop on Explainable, Transparent Autonomous Agents and Multi-Agent Systems* (pp. 39–53). Cham: Springer International Publishing.

Amann, J., Blasimme, A., Vayena, E., Frey, D., & Madai, V. I. (2020). Explainability for artificial intelligence in healthcare: A multidisciplinary perspective. *BMC Medical Informatics and Decision Making*, 20(1). https://doi.org/10.1186/s12911-020-01332-6

Arrieta, A. B., Díaz-Rodríguez, N., Del Ser, J., Bennetot, A., Tabik, S., Barbado, A., . . . Herrera, F. (2020). Explainable artificial intelligence (XAI): Concepts, taxonomies, opportunities and challenges toward responsible AI. *Information Fusion*, 58, 82–115.

Azumah, S. W., Elsayed, N., Adewopo, V., Zaghloul, Z. S., & Li, C. (2021, June). A deep lstm based approach for intrusion detection iot devices network in smart home. In *2021 IEEE 7th World Forum on Internet of Things (WF-IoT)*, IEEE, pp. 836–841.

Bedi, P., Goyal, S. B., Rajawat, A. S., Shaw, R. N., & Ghosh, A. (2022). Application of AI/IoT for smart renewable energy management in smart cities. *AI and IoT for Smart City Applications*, 115–138.

Chavhan, S., Gupta, D., Gochhayat, S. P., N, C. B., Khanna, A., Shankar, K., & Rodrigues, J. J. (2022). Edge computing AI-IoT integrated energy-efficient intelligent transportation system for smart cities. *ACM Transactions on Internet Technology*, 22(4), 1–18.

Čolaković, A., & Hadžialić, M. (2018). Internet of things (IoT): A review of enabling technologies, challenges, and open research issues. *Computer Networks*, 144, 17–39.

Feng, X., Zhu, X., Han, Q. L., Zhou, W., Wen, S., & Xiang, Y. (2022). Detecting vulnerability on IoT device firmware: A survey. *IEEE/CAA Journal of AutomaticaSinica*, 10(1), 25–41.

Fizza, K., Banerjee, A., Jayaraman, P. P., Auluck, N., Ranjan, R., Mitra, K., & Georgakopoulos, D. (2022). A survey on evaluating the quality of autonomic internet of things applications. *IEEE Communications Surveys & Tutorials*. 10.1109/COMST.2022.3205377. 13 September 2022.

Fotia, L., Delicato, F., & Fortino, G. (2023). Trust in edge-based internet of things architectures: State of the art and research challenges. *ACM Computing Surveys*, 55(9), 1–34.

Gruszka, A., Jupp, J. R., & De Valence, G. (2017). *Digital Foundations: How Technology is Transforming Australia's Construction Sector.*

Gupta, D., Bhatt, S., Gupta, M., & Tosun, A. S. (2021). Future smart connected communities to fight Covid-19 outbreak. *Internet of Things*, 13, 100342.

Humayun, M., Jhanjhi, N. Z., Alsayat, A., & Ponnusamy, V. (2021). Internet of things and ransomware: Evolution, mitigation and prevention. *Egyptian Informatics Journal*, 22(1), 105–117.

Istiaque Ahmed, K., Tahir, M., HadiHabaebi, M., Lun Lau, S., & Ahad, A. (2021). Machine learning for authentication and authorization in IoT: Taxonomy, challenges and future research direction. *Sensors*, 21(15), 5122.

Jagatheesaperumal, S. K., Pham, Q. V., Ruby, R., Yang, Z., Xu, C., & Zhang, Z. (2022). Explainable AI over the internet of things (IoT): Overview, state-of-the-art and future directions. *IEEE Open Journal of the Communications Society*. January, 2022.10.1109/OJCOMS.2022.3215676

Jamil, F., Ahmad, S., Iqbal, N., & Kim, D. H. (2020). Towards a remote monitoring of patient vital signs based on IoT-based blockchain integrity management platforms in smart hospitals. *Sensors*, 20(8), 2195.

Landgrebe, J. (2022). Certifiable ai. *Applied Sciences*, 12(3), 1050.

Lipton, Z. C. (2018). The mythos of model interpretability: In machine learning, the concept of interpretability is both important and slippery. *Queue*, 16(3), 31–57.

Macha, D., Kozielski, M., Wróbel, Ł., & Sikora, M. (2022). RuleXAI—A package for rule-based explanations of machine learning model. *SoftwareX*, 20, 101209.

Maglić, L., Grbčić, A., Maglić, L., & Gundić, A. (2021). Application of smart technologies in Croatian Marinas. *Transactions on Maritime Science*, 10(1), 178–188.

Mallik, A. (2019). Man-in-the-middle-attack: Understanding in simple words. *Cyberspace: Jurnal Pendidikan Teknologi Informasi*, 2(2), 109–134.

McCaig, M., Rezania, D., & Dara, R. (2023). Framing the response to IoT in agriculture: A discourse analysis. *Agricultural Systems*, 204, 103557.

Mohanta, B. K., Jena, D., Satapathy, U., & Patnaik, S. (2020). Survey on IoT security: Challenges and solution using machine learning, artificial intelligence and blockchain technology. *Internet of Things*, 11, 100227.

Mohseni, S., Zarei, N., & Ragan, E. D. (2021). A multidisciplinary survey and framework for design and evaluation of explainable AI systems. *ACM Transactions on Interactive Intelligent Systems (TiiS)*, 11(3–4), 1–45.

Muhsen, I. N., Rasheed, O. W., Habib, E. A., Alsaad, R. K., Maghrabi, M. K., Rahman, M. A., . . . Hashmi, S. K. (2021). Current status and future perspectives on the internet of things in oncology. *Hematology/Oncology and Stem Cell Therapy*. Volume 16, Issue 2, Article 2. https://doi.org/10.1016/j.hemonc.2021.09.003

Nwakanma, C. I., Ahakonye, L. A. C., Njoku, J. N., Odirichukwu, J. C., Okolie, S. A., Uzondu, C., . . . Kim, D. S. (2023). Explainable artificial intelligence (XAI) for intrusion detection and mitigation in intelligent connected vehicles: A review. *Applied Sciences*, 13(3), 1252.

Rathore, M. S., Poongodi, M., Saurabh, P., Lilhore, U. K., Bourouis, S., Alhakami, W., . . . Hamdi, M. (2022). A novel trust-based security and privacy model for internet of vehicles using encryption and steganography. *Computers and Electrical Engineering*, 102, 108205.

Rawal, A., McCoy, J., Rawat, D. B., Sadler, B. M., & Amant, R. S. (2021). Recent advances in trustworthy explainable artificial intelligence: Status, challenges, and perspectives. *IEEE Transactions on Artificial Intelligence*, 3(6), 852–866.

Rehman, A., Saba, T., Kashif, M., Fati, S. M., Bahaj, S. A., & Chaudhry, H. (2022). A revisit of internet of things technologies for monitoring and control strategies in smart agriculture. *Agronomy*, 12(1), 127.

Rejeb, A., Keogh, J. G., & Treiblmaier, H. (2019). Leveraging the internet of things and blockchain technology in supply chain management. *Future Internet*, 11(7), 161.

Rejeb, A., Rejeb, K., Treiblmaier, H., Appolloni, A., Alghamdi, S., Alhasawi, Y., & Iranmanesh, M. (2023). The internet of things (IoT) in healthcare: Taking stock and moving forward. *Internet of Things*, 100721.

Rojek, I., Jasiulewicz-Kaczmarek, M., Piechowski, M., & Mikołajewski, D. (2023). An artificial intelligence approach for improving maintenance to supervise machine failures and support their repair. *Applied Sciences*, 13(8), 4971.

Saba, T., Rehman, A., Sadad, T., Kolivand, H., & Bahaj, S. A. (2022). Anomaly-based intrusion detection system for IoT networks through deep learning model. *Computers and Electrical Engineering*, 99, 107810.

Sarker, I. H., Khan, A. I., Abushark, Y. B., & Alsolami, F. (2022). Internet of things (IoT) security intelligence: A comprehensive overview, machine learning solutions and research directions. *Mobile Networks and Applications*, 1–17.

Singh, S. K., Rathore, S., & Park, J. H. (2020). Blockiotintelligence: A blockchain-enabled intelligent IoT architecture with artificial intelligence. *Future Generation Computer Systems*, 110, 721–743.

Sinha, A., Garcia, D. W., Kumar, B., & Banerjee, P. (2023). Application of big data analytics and internet of medical things (IoMT) in healthcare with view of explainable artificial intelligence: A survey. In *Interpretable Cognitive Internet of Things for Healthcare* (pp. 129–163). Cham: Springer International Publishing.

Suwardi, I. S., & Surendro, K. (2014, November). An overview of multi agent system approach in knowledge management model. In *2014 International Conference on Information Technology Systems and Innovation (ICITSI)*, IEEE, pp. 62–69.

Vanino, S., Nino, P., De Michele, C., Bolognesi, S. F., D'Urso, G., Di Bene, C., . . . Napoli, R. (2018). Capability of Sentinel-2 data for estimating maximum evapotranspiration and irrigation requirements for tomato crop in Central Italy. *Remote Sensing of Environment*, 215, 452–470.

Yang, X., Shu, L., Liu, Y., Hancke, G. P., Ferrag, M. A., & Huang, K. (2022). Physical security and safety of IoT equipment: A survey of recent advances and opportunities. *IEEE Transactions on Industrial Informatics*, 18(7), 4319–4330.

Zolanvari, M., Yang, Z., Khan, K., Jain, R., & Meskin, N. (2021). Trust XAI: Model-agnostic explanations for ai with a case study on IIoT security. *IEEE Internet of Things Journal*, pp. 2967–2978. 10.1109/JIOT.2021.3122019.

Zouai, M., Kazar, O., Haba, B., & Saouli, H. (2017, December). Smart house simulation based multi-agent system and internet of things. In *2017 International Conference on Mathematics and Information Technology (ICMIT)*, IEEE, pp. 201–203.

Chapter 8

Application of Body Sensor Networks in Health Care Using the Internet of Medical Things (IoMT)

Salini Suresh, Srivatsala V., T. Kohila Kanagalakshmi, Suneetha V., Ganapati Hegde, and Hithesh R.

8.1 INTRODUCTION

Body sensor networks (BSNs) are a type of wearable technology that consist of various miniaturized sensors embedded or attached to a person's body to monitor and collect physiological and biomedical data. BSNs integrate the technology of sensing, intelligent information processing, pervasive computing, and communication. Body sensor devices process information and provide a near-accurate view of the physiological, behavioral, health, or emotional state of the person wearing it. By using wearable sensors to collect data from various body parts, these networks enable medical professionals to better understand how the body functions and what is happening inside it. These networks are designed to measure specific health-related parameters and can be used for a wide range of applications, including health care, sports, wellness, and research.

8.1.1 Typical Key Components of a Body Sensor Network

8.1.1.1 Sensors

These are the primary elements of BSNs and are responsible for capturing various physiological data from the wearer. Common types of sensors used in BSNs include the following:

Electrocardiogram (ECG) sensors: Measure heart activity.
Accelerometers: Measure movement and orientation.
Temperature sensors: Measure body temperature.
Blood oxygen sensors: Measure oxygen saturation in the bloodstream.

8.1.1.2 Sensor Nodes

A sensor node refers to a device that incorporates one or more sensors, along with processing capabilities, storage, and communication interfaces.

DOI: 10.1201/9781003406723-8

Sensor nodes are responsible for capturing data from the sensors, processing and analyzing the data, extracting features, and transmitting relevant information to a central entity (sink node or gateway) in the network.

8.1.1.3 Communication Module

BSNs utilize wireless communication protocols to transmit data from the sensor nodes to a central data collection point. Depending on the range and data rate requirements, these networks can employ various wireless technologies such as Bluetooth, Zigbee, Wi-Fi, and even cellular networks.

8.1.1.4 Power Supply

As BSNs are typically worn on the body, they need to be powered by a reliable and compact energy source such as rechargeable batteries and energy-harvesting mechanisms.

8.1.1.5 User Interface

In some cases, BSNs may have a user interface that provides feedback to the wearer or enables interaction with the device, such as a display, LED indicators, and haptic feedback.

8.2 CATEGORIES OF BODY SENSORS

Body sensors, often termed as wearable or implantable sensors, represent a revolutionary integration of technology and medicine. These tiny devices, whether worn on the surface, sewn into clothing, or even inserted within the body, continually monitor and capture a plethora of physiological and biomechanical data. The different categories of body sensors are listed as follows:

Cardiac and Cardiovascular Sensors: Electrocardiogram (ECG/EKG) sensors measure the electrical activity of the heart. ECG sensors detect the electrical impulses generated by the polarization and depolarization of cardiac tissue. Electrodes are placed on the skin, and the voltage difference between electrodes is measured, producing an ECG waveform.

Photoplethysmogram (PPG) Sensors: Detect blood volume changes in the microvascular bed of tissue used to measure heart rate, heart rate variability, and sometimes blood oxygen saturation. PPG sensors use LEDs to shine light (usually green) into the skin. The light reflected back varies due to blood flow changes. By analyzing these variations, the sensors determine heart rate and other vascular parameters.

Blood Pressure Sensors: Monitor arterial blood pressure.

8.2.1 Respiratory Sensors

Spirometers: Measure the volume of air inhaled and exhaled by the lungs. When the person breathes into the device, it measures the volume and flow of air during inhalation and exhalation, providing insights into lung function.

Respiration Rate Sensors: Monitor the frequency of inhalation and exhalation.

Capnography Sensors: Measure the concentration of carbon dioxide in exhaled air.

Electroencephalogram (EEG) Sensors: Monitor the electrical activity of the brain.

Electromyogram (EMG) Sensors: Detect the electrical activity of muscles. EMG sensors use electrodes to detect electrical potentials generated by muscle cells when they are neurologically activated. The detected signals can be analyzed to determine muscle activity.

8.2.2 Oximetry

Pulse Oximeters: Measure the percentage of oxygen-saturated hemoglobin (SpO2) in the blood. These devices use LEDs to emit two different wavelengths of light (usually red and infrared) through a fingertip or earlobe. By measuring the absorption of each wavelength, the device calculates the percentage of oxygenated hemoglobin in the blood.

8.2.3 Thermal Sensors

Body Temperature Sensors: Often appear in the form of wearable patches or ingestible pills to measure core body temperature. Thermistors are resistors with resistance values that change significantly with temperature. Thermocouples produce a voltage based on the temperature difference between two junctions. Both can be used to determine body temperature.

8.2.4 Movement and Posture Sensors

Accelerometers: Measure acceleration forces, which can indicate movement, orientation, and activity levels. Accelerometers detect changes in motion and orientation using tiny capacitive plates inside the sensor. Movement causes shifts in these plates, which results in voltage changes that are processed to determine motion.

Gyroscope Sensors: Measure angular velocity and can determine orientation and rotation.

Magnetometers: Measure magnetic fields, often used in combination with accelerometers and gyroscopes for orientation and movement detection.

8.2.5 Biochemical Sensors

Glucose Sensors: Continuously or intermittently monitor glucose levels in the body, often used by diabetics. For glucometers, a blood sample is applied to a test strip, which uses an enzyme reaction to produce an electric current proportional to glucose concentration. Continuous glucose monitors (CGMs) use a subcutaneous sensor to measure glucose in interstitial fluid.

Lactate Sensors: Measure lactate concentration, which can be an indicator of muscle fatigue.

Electrolyte Sensors: Monitor levels of specific ions such as potassium or sodium in the body.

Hydration and Sweat Analysis Sensors: Measure the hydration level or analyze the chemical composition of sweat to determine various physiological parameters.

8.2.6 Acoustic Sensors

Stethoscopes and Phonocardiography Sensors: Listen to heartbeats and other body sounds.

Pulmonary Sound Sensors: Monitor lung sounds to detect issues like wheezing or crackling.

8.2.7 Optical and Imaging Sensors

Dermatoscopes: Used for skin examinations.

Retinal Imaging Devices: Monitor the health of the eyes.

8.2.8 Bioimpedance Sensors

Bioimpedance sensors measure body composition and fluid shifts. A small electrical current is passed through the body, and the impedance or resistance is measured. Since different tissues (e.g., fat, muscle) have different resistance values, this can be used to estimate body composition.

8.2.9 Environmental Sensors

These are sometimes combined with body sensors in wearables.

UV Sensors: Monitor ultraviolet exposure from the sun.

Ambient Temperature and Humidity Sensors: Monitor the surrounding environment.

8.3 ROLE OF BODY SENSOR NETWORKS IN HEALTH CARE

BSNs have significant applications and potential in health care. These networks can play a crucial role in improving patient monitoring, diagnostics, treatment, and overall healthcare management. The following are some specific ways BSNs are utilized in health care.

8.3.1 Remote Patient Monitoring

BSNs enable continuous monitoring of patients' vital signs and health parameters, even when they are outside healthcare facilities. This is particularly beneficial for patients with chronic conditions, in post-operative care, or requiring long-term monitoring. Remote monitoring allows healthcare providers to detect early signs of deterioration, provide timely interventions, and reduce hospital readmissions.

8.3.2 Telemedicine and Telehealth

BSNs can be integrated into telemedicine platforms, facilitating virtual consultations between patients and healthcare professionals. The real-time data captured by BSNs can be transmitted to healthcare providers, enabling them to assess the patient's condition remotely and make informed decisions about treatment plans. Data concerning physical activity, medication adherence, dietary habits, exercise routines, and stress levels provide users with valuable insights into their overall health and well-being. This data can assist healthcare providers in offering personalized advice and support to improve patients' lifestyle choices.

8.3.3 Chronic Disease Management

BSNs play a vital role in managing chronic diseases like diabetes, hypertension, and cardiovascular conditions. By continuously monitoring relevant health metrics, healthcare providers can adjust treatment plans and medications as needed to maintain patients' health within target ranges.

8.3.4 Fall Detection and Elderly Care

BSNs equipped with accelerometers and other motion sensors can detect falls in elderly individuals or patients with mobility issues. These systems can automatically alert caregivers or emergency services, ensuring prompt assistance in case of accidents or emergencies.

8.3.5 Post-Operative Monitoring

After surgery, BSNs can be used to monitor patients' vital signs and recovery progress. This allows healthcare providers to identify any postoperative complications early and provide appropriate care.

8.3.6 Clinical Research and Trials

BSNs are valuable tools in clinical research and trials, providing objective and continuous data on patients' health status and treatment responses. This data can lead to better understanding and optimization of medical treatments.

8.4 BENEFITS OF BODY SENSOR NETWORKS IN HEALTH CARE

BSNs offer numerous benefits in health care, revolutionizing the way health care is delivered, monitored, and managed. Some of the key advantages of BSNs in health care include the following:

8.4.1 Continuous Monitoring

BSNs enable continuous and real-time monitoring of patient's vital signs and health parameters, providing a more comprehensive view of their health status. This continuous monitoring allows healthcare providers to detect subtle changes or early signs of deterioration, facilitating timely interventions and personalized care.

8.4.2 Timely Interventions

With BSNs, healthcare providers can receive alerts and notifications in real time when a patient's health parameters deviate from normal ranges or when critical events occur. This enables immediate responses and interventions, preventing potential complications and emergencies.

8.4.3 Personalized Health Care

BSNs provide detailed and individualized data about patients' health, allowing healthcare providers to tailor treatment plans and interventions based on each patient's unique needs and responses. This personalized approach can lead to more effective treatments and improved patient satisfaction.

8.4.4 Enhanced Data Accuracy

BSNs provide objective and accurate data, minimizing human errors and subjectivity in data collection. The high-quality data obtained from BSNs can improve the accuracy of diagnoses and treatment decisions.

8.4.5 Remote Consultations

BSNs integrated with telemedicine platforms facilitate remote consultations between patients and healthcare professionals. This not only improves access to healthcare services, especially for patients in remote areas, but also allows for more frequent follow-ups and better patient engagement.

8.4.6 Data-Driven Insights

BSNs generate vast amounts of health data that can be analyzed and used to derive valuable insights into patient health trends, treatment outcomes, and population health patterns. This data-driven approach can lead to evidence-based improvements in healthcare practices and policies.

8.4.7 Early Disease Detection

BSNs enable early detection of health issues and disease progression, allowing for prompt medical attention and potential prevention of complications.

8.4.8 Improving Lifestyle Behaviors

BSNs can track patients' lifestyle behaviors such as physical activity, sleep patterns, and dietary habits. This data can be used to promote healthier behaviors and encourage patients to adopt lifestyle changes that positively impact their health.

8.4.9 Research and Innovation

BSNs contribute to advancements in medical research and innovation by providing researchers with real-world data from diverse populations. This data can be used to develop and refine medical technologies, treatments, and preventive strategies.

Overall, BSNs have the potential to significantly improve healthcare outcomes, enhance patient experiences, and promote more proactive and patient-centered healthcare delivery. As the technology continues to evolve, BSNs are expected to play an increasingly vital role in transforming the healthcare landscape.

8.5 IoT AND IoMT

8.5.1 What Is IoMT?

Internet of Medical Things (IoMT) refers to medical devices and applications with internet connectivity. IoMT refers to the integration of medical devices, sensors, and healthcare systems with IoT. It involves connecting medical devices and equipment to networks and the internet to enable data collection, analysis, and communication for improved healthcare delivery. We also refer to it as the "Internet of Things in health care."

The overall category of IoT devices is typically more consumer-oriented, focusing on usability and convenience. IoT devices include smart TVs, lighting apps, voice assistants—really any number of smart, connected devices. IoMT devices and applications are designed with health care in mind, including the following:

Smart thermometers and infusion pumps
Remote patient monitoring (R\PM) devices
Personal emergency response systems (PERS)
Heart rate sensors and glucose monitors
Ingestible sensors and cameras
MRI machines

8.5.2 Importance of IoMT

Healthcare professionals can provide quicker and better care thanks to connected medical equipment. From robotic surgery to glucose monitoring, there are many applications. IoMT advantages include the following:

- Enhanced therapies and financial savings.
- Faster and more accurate diagnosis because of IoMT technology's ability to track patients' vital signs in great detail.
- Better patient monitoring without the need for trips to a hospital.

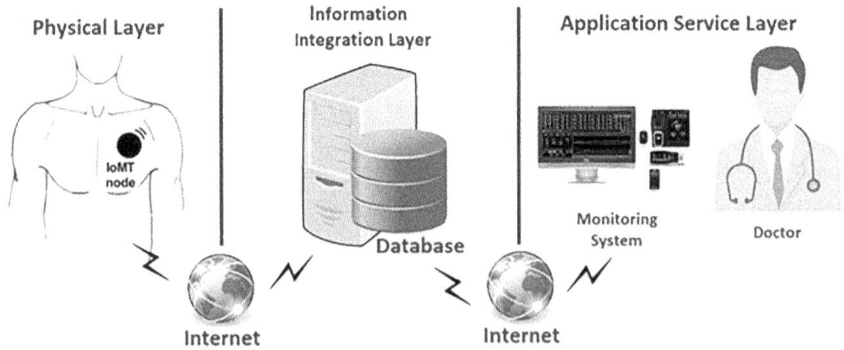

Figure 8.1 IoMT architecture overview [1].

8.5.3 IoMT Architecture Overview

Medical equipment, healthcare systems, and data analytics platforms can all be seamlessly connected and integrated thanks to IoMT architecture. It is intended to assist the secure and interoperable gathering, management, and analysis of healthcare data.

Devices and sensors, communication, and data management make up the key elements of the IoMT architecture.

Devices and Sensors: The IoMT architecture includes a range of medical devices and sensors, such as remote patient monitors, wearable health trackers, and diagnostic tools. These gadgets gather live health information like vital signs, activity levels, and medication compliance.

Connectivity: The IoMT architecture depends on a number of connection technologies, including Wi-Fi, Bluetooth, cellular networks, and specialized medical device communication protocols, to make data transmission easier.

Data Management: To store, process, and manage the enormous volume of healthcare data produced by the linked devices, the IoMT design makes use of cloud infrastructure. Scalability, dependability, and safe data storage are provided via cloud servers, databases, and computer resources. The gathered healthcare data is processed and analyzed using data analytics platforms and machine learning.

The IoMT architecture places a premium on security and privacy. To prevent unauthorized access to sensitive patient information, strong security mechanisms including encryption, authentication, and access control are put in place. Healthcare data is managed and shared securely thanks to privacy laws and regulatory requirements.

In order to provide seamless communication and data exchange between various healthcare systems and devices, interoperability is also crucial to the IoMT architecture. Interoperability between various IoMT components is guaranteed by standardized communication protocols and health data exchange standards.

Applications and user interfaces allow patients and healthcare professionals a way to communicate with the IoMT system. Real-time monitoring, data visualization, and remote healthcare management are made possible through mobile apps, web portals, and specialized healthcare software.

Overall, the IoMT architecture transforms healthcare delivery by utilizing cloud computing, networked devices, and data analytics. Improved patient outcomes, personalized treatment, preventive care, and remote patient monitoring are all made possible. Healthcare providers can improve efficiency, accuracy, and patient-centricity in healthcare services by utilizing the IoMT architecture's power.

8.6 WEARABLE BSNs

Wearable BSNs refer to a collection of wearable sensors and devices that are integrated into a network to monitor and gather physiological, biomechanical, and environmental data from the human body. BSNs have found applications in various fields, including health care, sports, wellness, and research. These networks enable real-time monitoring, analysis, and interpretation of a person's physiological parameters, allowing for insights into their health status, activity levels, and overall well-being.

Smartwatches and Fitness Trackers: These are perhaps the most common types of wearable BSNs. They typically include sensors such as accelerometers, heart rate monitors, and sometimes even GPS. Smartwatches and fitness trackers are popular for monitoring activity levels, heart rate, sleep patterns, and overall fitness.

Wearable ECG Monitors: These devices monitor the electrical activity of the heart and can detect irregularities such as arrhythmias. They provide a portable and continuous ECG recording that can be useful for diagnosing and managing heart conditions.

Smart Clothing: Certain types of clothing are embedded with sensors to monitor various parameters. For example, smart shirts can measure breathing rate, body temperature, and posture. These garments are particularly useful for applications in sports, fitness, and health care.

Wearable Glucose Monitors: Designed primarily for individuals with diabetes, these devices measure blood glucose levels continuously and non-invasively, reducing the need for frequent finger stick tests.

Wearable Blood Pressure Monitors: These devices provide continuous or on-demand monitoring of blood pressure, allowing individuals to track their cardiovascular health throughout the day.

Smart Glasses: Equipped with sensors and displays, smart glasses can provide real-time data on environmental conditions, gaze tracking, and even augmented reality overlays for various professional and medical applications.

Wearable Pulse Oximeters: These devices measure blood oxygen saturation (SpO2) and pulse rate. They are especially relevant for monitoring respiratory health and can be valuable for individuals with conditions like sleep apnea or lung diseases.

Wearable Temperature Monitors: These devices track body temperature variations, which can be useful for monitoring fever or detecting temperature changes related to ovulation in women.

Wearable EEG Monitors: Electroencephalogram sensors in wearable devices can measure brain activity. These devices have applications in neurology, sleep monitoring, and brain-computer interfaces.

Wearable Motion Sensors: These sensors track movement and orientation. They are commonly used for gait analysis, fall detection, and rehabilitation monitoring.

Wearable Skin Sensors: Sensors that can detect sweat composition, skin temperature, and other skin-related metrics offer insights into hydration levels, stress, and physical exertion.

Wearable Respiratory Monitors: These devices track breathing patterns, chest movements, and lung function, providing valuable data for respiratory conditions and wellness.

The diversity of wearable BSN devices reflects the growing range of applications and the potential to monitor and improve various aspects of human health and performance. Each type of wearable BSN serves a specific purpose, and their combined use can provide a comprehensive picture of an individual's well-being.

8.6.1 Wearable BSN Example

For guiding and supporting athletes, smart sensors are being embedded in clothing and accessories. A prominent example is the adidas_1 running shoe, which is the first shoe that features a wearable embedded system. This shoe is built to adapt to various running conditions like the prevailing surface situation or the runner's speed and fatigue state by changing the cushioning of the sole using a motor driven cable system.

Figure 8.2 Hexoskin biometric shirt [2].

8.6.2 Wearable BSN-Based ECG-Recording System

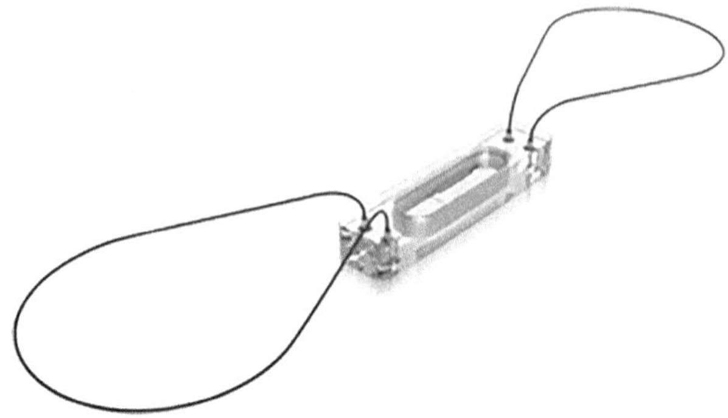

Figure 8.3 Wearable BSN-based ECG recording system [3].

8.6.3 Hexoskin Biometric Shirt

Hexoskin Smart Garments continuously monitor the user's cardiorespiratory function, sleep, and activity with textile-embedded sensors. The garments are comfortable, non-invasive clothing, widely tested and adopted by thousands of clients and patients since 2013.

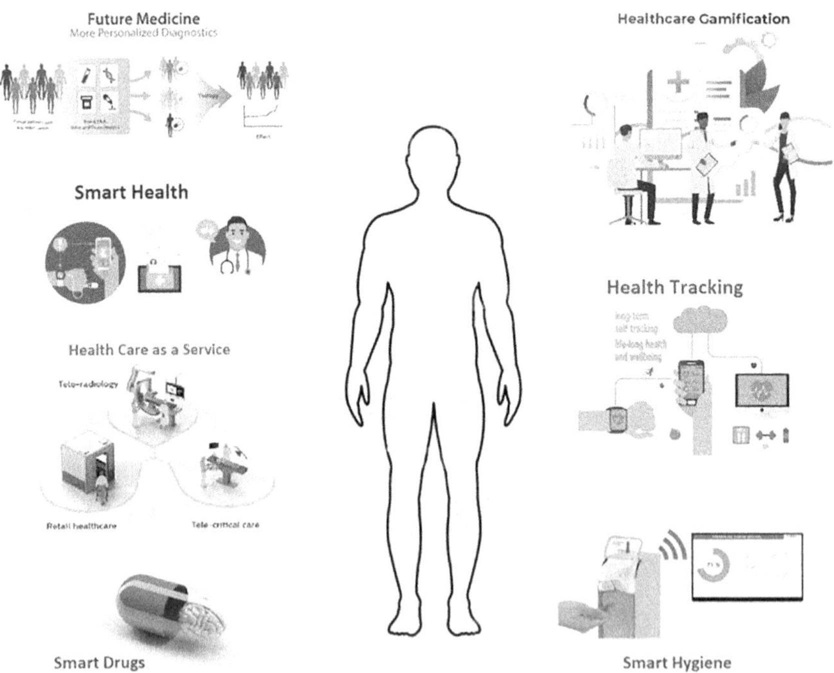

Figure 8.4 Hexoskin biometric shirt [4].

8.7 IMPLANTABLE BSN

An implantable BSN is a network of miniaturized sensors and devices that are surgically implanted within the human body to monitor various physiological parameters and transmit data wirelessly for analysis and interpretation. Unlike wearable BSNs that are worn on the body's surface, implantable BSNs are placed beneath the skin or within specific organs or tissues. These networks are designed to provide continuous and real-time monitoring of internal body functions, enabling healthcare professionals to gather insights into a patient's health status, detect abnormalities, and make informed medical decisions.

Implantable Sensors: These are small, biocompatible devices equipped with sensors that can measure a range of physiological parameters such as temperature, pressure, glucose levels, pH, oxygen saturation, and more. These sensors are specifically designed to withstand the internal environment of the body.

Implantation Procedure: Implantable BSN devices are typically inserted into the body through minimally invasive surgical procedures. The location of implantation depends on the specific parameter being monitored. For example, glucose sensors may be implanted in the subcutaneous tissue, while cardiac sensors may be placed within the heart.

Wireless Communication: Just like wearable BSNs, implantable sensors within a network communicate wirelessly, using technologies such as radiofrequency (RF) communication or near-field communication (NFC). These sensors transmit data to external receivers or devices for further processing and analysis.

Data Collection and Transmission: The implanted sensors continuously gather data from inside the body and transmit it wirelessly to external devices or data storage centers. This allows healthcare professionals to remotely monitor the patient's health status without the need for physical connections.

Data Analysis and Interpretation: The collected data is processed and analyzed using advanced algorithms and machine learning techniques. Patterns, trends, and anomalies are identified to provide insights into the patient's health condition and to trigger alerts if critical thresholds are exceeded.

Real-Time Monitoring and Medical Interventions: Implantable BSNs enable real-time monitoring of a patient's physiological parameters. This can lead to early detection of medical issues, allowing for timely interventions and adjustments to treatment plans.

8.7.1 Examples

CardioMEMS: Developed implantable pressure sensors that can measure pressure after being implanted into an aneurysm sac during an endovascular repair.

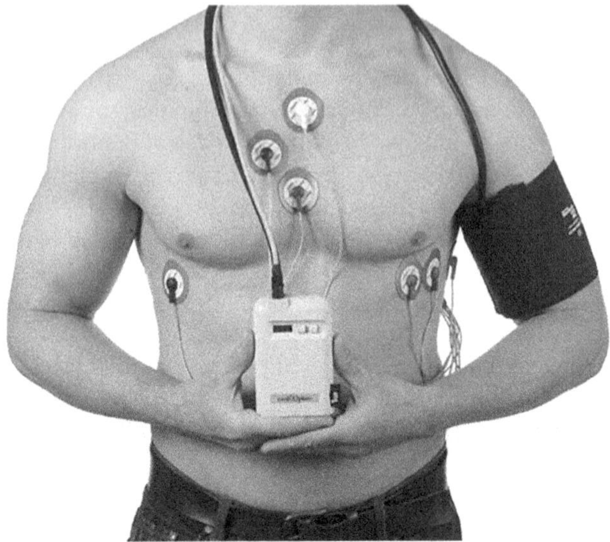

Figure 8.5 CardioMEMS [5].

8.8 IMPLANTABLE BIOSENSOR CHIP

The implantable biosensor chip developed by EPFL Integrated Systems Laboratory is implanted on the skin surface. It is connected to a smartphone to measure multiple biochemistry indicators at the same time. This helps keep track of cholesterol, glucose, and drug concentrations.

8.9 TYPES OF IMPLANTABLE BSN

Implantable BSNs consist of miniature sensors and devices that are surgically implanted inside the human body to monitor various physiological parameters and transmit data for analysis. These implantable devices are used in health care, research, and medical interventions. The following are some types of implantable BSN devices:

Implantable Cardiovascular Monitors: These devices are placed within the cardiovascular system to monitor parameters such as heart rate, blood pressure, and cardiac electrical activity (ECG). They can be used for long-term monitoring of heart conditions and arrhythmias.

Implantable Glucose Monitors: These devices continuously measure blood glucose levels from within the body, providing real-time data for individuals with diabetes. They reduce the need for frequent finger stick tests.

Implantable Neurological Monitors: These devices are used to monitor brain activity, intracranial pressure, and other neurological parameters. They are valuable for diagnosing and treating conditions like epilepsy and traumatic brain injuries.

Implantable Temperature Sensors: These sensors monitor internal body temperature, helping detect fever or changes in body temperature for medical diagnoses.

Implantable Drug Delivery Systems: These devices deliver medications or therapeutic substances directly into the body, providing controlled and targeted treatment for conditions like chronic pain, Parkinson's disease, or cancer.

Implantable Pacemakers: These devices are used to regulate abnormal heart rhythms by sending electrical impulses to the heart muscles, ensuring proper heart function.

Implantable Defibrillators: These devices monitor heart rhythms and deliver electrical shocks when life-threatening arrhythmias are detected to restore a normal heart rhythm.

Implantable Neural Interfaces: These devices establish a direct communication link between the brain or nervous system and external devices, enabling prosthetics control, restoring movement in paralysis cases, and assisting research in neurology.

Implantable Blood Pressure Sensors: These sensors monitor blood pressure from within the body, providing continuous and accurate measurements for cardiovascular health management.

Implantable Bone Growth Stimulators: These devices use electrical or electromagnetic fields to promote bone healing and growth; they are particularly useful in orthopedic and fracture cases.

Implantable Electrodes for Deep Brain Stimulation (DBS): DBS electrodes are implanted in the brain to treat conditions like Parkinson's disease, tremors, and mood disorders by delivering electrical pulses to specific brain areas.

Implantable Insulin Pumps: These devices automatically deliver insulin to individuals with diabetes, mimicking the natural insulin release of a healthy pancreas.

Implantable RFID Tags: Radio frequency identification (RFID) tags can be implanted for patient identification, medical record retrieval, and ensuring accurate treatments in medical settings.

Implantable Telemetry Devices: These devices monitor various physiological parameters and wirelessly transmit the data to external receivers for real-time monitoring and analysis.

Each type of implantable BSN serves a specific medical purpose, and the field continues to advance with ongoing research and development. These devices play a critical role in improving patient care, providing accurate data for medical decisions, and advancing our understanding of the human body.

8.10 BSN IN IoMT

BSNs can be used for remote patient monitoring, chronic disease management (e.g., diabetes, cardiovascular conditions), and post-operative care. Athletes can use BSNs to track their physical activity, monitor vital signs, optimize training routines, and prevent injuries. BSNs can provide insights into sleep patterns, stress levels, and overall wellness. Researchers can gather large-scale data for studies on human behavior, health trends, and disease patterns.

8.10.1 IoMT-Based WBSN Framework for Healthcare Monitoring System

This term refers to the integration and usage of BSN health care to monitor the health of humankind by using the IoT technology, which in turn may be called IoMT. The framework for a healthcare monitoring system refers to an organized structure or approach for designing, implementing, and managing a system that monitors various aspects of a person's health. It involves the integration of technology, processes, and data management to achieve effective healthcare monitoring.

Various components of such a framework include wearables attached to the human body to collect blood pressure, sugar levels, heart rate, body temperature, and so on by using various sensors and thus continuously monitoring all the aforementioned parameters. The collection of data is an essential feature of this framework for further analysis. Wireless communication plays a crucial role in this context, given that data is collected and recorded remotely. Securing the data of the patient is also a major part of the framework. Also, it is essential to keep the patient informed through various alerts and notifications and gain valuable insights into the patient's healthy future.

The IoMT-based wireless BSN (WBSN) framework for healthcare monitoring systems holds significant promise for remote patient monitoring, chronic disease management, preventive care, and overall health and wellness. It aligns with the broader trend of leveraging technology to enhance healthcare outcomes and empower individuals to take proactive control of their health.

8.11 APPLICATIONS

8.11.1 Disease Prediction

Internet-assisted disease prediction is a cutting-edge approach to early identification, close monitoring, and quick treatment based on software applications and trained artificial intelligence models or electronic devices for the creation of next-generation healthcare innovation. The cornerstone and skeleton of the entire medical industry is now the use of IoT and the electronic web in healthcare institutions. It can carry out a variety of tasks, including monitoring a patient's health, analyzing a patient's medical history, monitoring digital trails, keeping an eye on drugs, and organizing equipment and patient health care. It can also be used to monitor a patient's medication intake and use alert systems.

Point of care (POC) diagnostics are used for bedside medical testing with rapid findings that improve the speed of the treatment process and also assist patients who are elderly, are physically challenged, have chronic conditions, and are in need of immediate attention. For instance, the development of handheld portable ultrasound scanners (PUS) has been made possible by the use of field programmable gate arrays (FPGA), digital signal processors (DSP), and graphics processors in ultrasound technology. POC diagnosis, remote health care, and emergency scenarios are made simple by this.

The use of telesonography technology using high-efficiency video coding (HEVC) and H.264 encoding, in which a non-expert conducts POC ultrasonography and is promptly forwarded to an expert for diagnosis, could help to alleviate the scarcity of expert sonographers. A variety of healthcare solutions are offered by Comarch Healthcare. With expertise in IoT, artificial intelligence, cloud platforms, m-Health, and cybersecurity for the healthcare business, as well as software products for radiography, remote medical care, and medical record administration, and hospital IT tools.

Disease prediction tools:

- *Comarch Diagnostic Point*: This addresses the problems that a medical facility is trying to solve, such as appointment availability and scheduling, time management, and cost management. It is made up of tools and software that make it simple and quick to measure each patient's fundamental vital signs. It works with a tablet app that uses Bluetooth to gather information from the patient's peripheral device and send it to a remote care center.
- *Comarch Life Wristband*: Patients can speak with one other and request assistance from the telehealth center using these always wearable, waterproof devices with lengthy battery lives. The sensors automatically warn a telecare center after identifying loss of consciousness. The medical staff has access to all of a patient's personal and emergency data as well as their electronic health records (EHR).
- *Comarch CardioVest*: This is made to diagnose, provide prophylactic diagnosis, and monitor adult cardiac patients. It captures and sends ECG data to a telemedicine platform, which analyzes it and conducts a preliminary study into the circumstance and divergence from the norm.

8.11.2 Non-Invasive E-Medical Care

Non-invasive e-medical care in the context of IoT refers to the use of IoT technologies to monitor, diagnose, and treat medical conditions without the need for invasive procedures or physical contact with the patient. This approach leverages IoT devices, sensors, and data analytics to enable remote and continuous monitoring of a patient's health, providing timely interventions and personalized care.

The following are key aspects of non-invasive e-medical care in IoT:

- *Wearable Devices*: IoT-enabled wearables can gather a variety of physiological data, including heart rate, blood pressure, temperature, and activity levels. Examples include smartwatches, fitness trackers, and medical patches. These devices provide real-time health information and help monitor chronic conditions like diabetes, heart disease, and respiratory disorders without the need for invasive procedures.
- *Remote Patient Monitoring (RPM)*: IoT-based RPM systems allow healthcare providers to monitor patients remotely. Connected medical devices and sensors transmit data securely to healthcare facilities, enabling healthcare professionals to track a patient's health status and detect potential issues early on. This reduces the need for frequent hospital visits and improves patient outcomes.
- *Telemedicine and Telehealth*: IoT facilitates telemedicine and telehealth services by enabling remote consultations between patients and healthcare providers. Video conferencing and IoT devices allow doctors to assess patients' conditions, offer medical advice, and adjust treatment plans without requiring in-person visits.
- *Data Analytics and Machine Learning*: Data analytics and machine learning techniques are used to process and analyze the enormous volume of data produced by IoT devices. These innovations enable early intervention and individualized treatment regimens for patients by identifying patterns, forecasting health trends, and offering actionable insights.
- *Medication Management*: Patients can successfully manage their medications with the use of IoT-enabled smart pill dispensers and medication adherence trackers. These gadgets can remind patients to take their pills on time and alert medical staff if a patient forgets to do so, increasing medication compliance and treatment success.
- *Ambient Assisted Living*: IoT technologies can be integrated into the home environment to support elderly or disabled individuals. Smart home devices, such as motion sensors, fall detectors, and voice-activated assistants, can enhance safety and enable seniors to age in place comfortably.
- *Health Tracking Apps*: IoT-powered mobile applications allow patients to monitor their health, record symptoms, and track progress over time. These apps can also offer health tips and personalized recommendations based on the collected data.

One of the significant advantages of non-invasive e-medical care in IoT is that it enhances patient comfort and convenience while reducing the burden on healthcare facilities. It enables early detection of health issues, leading to timely interventions and better management of chronic conditions. However, ensuring data security, privacy, and interoperability between different IoT devices and platforms remain important challenges in the adoption of these technologies. Regulatory compliance and maintaining the accuracy and reliability of IoT-based medical devices are also critical factors for the successful implementation

The following are some important concepts about non-invasive e-medical care:

- *Chronic Disease Management*: IoT-enabled non-invasive e-medicine can significantly enhance the management of chronic conditions like diabetes, hypertension, and asthma. Continuous monitoring of relevant health parameters allows for proactive adjustments to treatment plans, reducing the risk of complications and hospitalizations.
- *Preventive Health Care*: IoT devices can facilitate preventive health care by monitoring lifestyle factors like physical activity, sleep patterns, and nutrition. Patients can receive personalized recommendations to maintain a healthier lifestyle, potentially reducing the risk of developing chronic conditions.
- *Real-Time Alerts and Notifications*: IoT devices can be programmed to send real-time alerts and notifications to healthcare providers and caregivers if abnormal health trends or emergencies are detected. This enables rapid responses and immediate medical attention, improving patient safety.
- *Post-Surgery Monitoring*: Non-invasive e-medical care in IoT can be valuable for post-surgery monitoring. IoT-enabled devices can track a patient's vital signs and recovery progress remotely, reducing the need for extended hospital stays and allowing for more flexible post-operative care plans.
- *Remote Rehabilitation*: IoT-based solutions can be utilized for remote rehabilitation programs, where patients can follow prescribed exercises at home under the guidance of virtual physiotherapists. This approach enhances accessibility to rehabilitation services and improves patient compliance.
- *Elderly Care*: IoT devices play a significant role in elderly care by providing continuous monitoring and support to seniors living independently. IoT technologies can detect falls, monitor daily activities, and alert caregivers in case of emergencies, ensuring the safety and well-being of the elderly population.
- *Mental Health Support*: IoT-powered wearables and mobile apps can aid in monitoring mental health conditions, such as stress and anxiety.

Patients can track their emotional states and receive timely interventions or coping strategies through the application.

- *Personalized Medicine*: IoT data, when combined with advanced analytics and artificial intelligence, enables the development of personalized medicine approaches. By analyzing a patient's specific health data, healthcare providers can tailor treatments and interventions to meet individual needs more effectively.
- *Remote Diagnostics*: IoT devices featuring medical-grade sensors can collect precise and continuous data, facilitating remote diagnostics. Healthcare professionals can remotely assess a patient's condition and provide diagnostic insights without the need for in-person examinations.
- *Emergency Response and Disaster Management*: In emergency situations and natural disasters, IoT-based medical devices can be used to quickly deploy medical resources and provide timely medical assistance to affected populations.
- *Clinical Trials and Research*: IoT technologies can improve the efficiency of clinical trials by collecting real-time data from participants and reducing the need for physical visits. This leads to faster data acquisition and more accurate insights during the research process.
- *Access to Health Care in Remote Areas*: Non-invasive e-medical care through IoT can extend medical services to remote and underserved areas. Patients in such regions can access telemedicine consultations and remote monitoring, improving healthcare accessibility and equity.

Overall, non-invasive e-medical care in IoT holds tremendous potential to transform healthcare delivery, making it more patient-centric, efficient, and accessible. As technology continues to advance, the integration of IoT in health care is expected to become more seamless and impactful, benefiting both patients and healthcare providers.

8.11.3 Ambient Assisted Living (AAL)

Ambient assisted living (AAL) in IoT refers to the use of IoT technologies to create a supportive and intelligent living environment for elderly or disabled individuals. By giving these people a secure, comfortable environment, AAL hopes to improve their quality of life and provide a comfortable and independent living experience through the seamless integration of smart devices and sensors within their living spaces.

The following are the key components and features of AAL in IoT:

- *Smart Home Devices*: AAL utilizes IoT-enabled smart thermostats, motion sensors, door/window sensors, smart lighting, and smart appliances to create an interconnected and automated living environment.

These devices can be controlled remotely through smartphones or voice-activated assistants.

- *Fall Detection and Prevention*: IoT sensors placed strategically throughout the living space can detect falls or abnormal movements. When a fall is detected, the system can automatically trigger an alert to caregivers or emergency services, enabling a prompt response.
- *Remote Monitoring*: IoT devices can monitor the daily activities and routines of individuals living alone or with health conditions. The data collected, such as movement patterns, sleep quality, and activity levels, can provide insights into their well-being and health status.
- *Medication Management*: AAL systems can incorporate smart pill dispensers or medication reminder devices to help individuals manage their medication schedules effectively. Alerts can be sent to remind them to take their medications on time.
- *Emergency Response*: AAL in IoT can facilitate emergency response systems, allowing individuals to call for help easily in case of medical emergencies. These systems can include panic buttons or voice-activated emergency call services.
- *Voice Assistants*: Virtual voice assistants like Google Assistant and Amazon Alexa are integrated into the AAL ecosystem, enabling voice-controlled operation of various devices and providing information or reminders.
- *Behavioral Analysis*: By analyzing data collected from IoT sensors, machine learning algorithms can identify patterns and detect deviations from normal behavior. This analysis can help identify potential health issues or safety concerns.
- *Social Interaction*: AAL systems can incorporate video conferencing or messaging capabilities to enable remote social interactions with family members, friends, or caregivers, reducing feelings of isolation.
- *Personalized Services*: AAL in IoT can adapt to an individual's preferences and needs over time, offering personalized services tailored to their specific requirements.
- *Privacy and Security*: Ensuring data privacy and security is of paramount importance in AAL systems. Data encryption, secure communication protocols, and access controls are used to safeguard sensitive information.
- *Aging in Place*: AAL enables elderly individuals to age in place by providing them with a safe and supportive living environment. This reduces the need for institutionalized care and promotes independent living.
- *Caregiver Support*: AAL in IoT can also assist caregivers by providing them with insights into the well-being of their loved ones and easing the burden of constant physical monitoring.

The integration of IoT technologies in AAL has the potential to revolutionize elderly care and support individuals with disabilities, promoting their independence, safety, and overall well-being. As technology continues to advance, AAL systems are anticipated to advance in sophistication and be seamless and user-friendly, further enhancing the living experience for those who benefit from these solutions.

8.11.4 Remote Health Monitoring

Using digital medical instruments like a weight scale, blood pressure monitor, pulse oximeter, and blood glucose meter, healthcare professionals monitor patients through a telehealth practice known as remote health monitoring.

Patients with both acute self-limited illnesses and chronic disease states use remote health monitoring, which includes glucose meters for diabetic individuals, blood pressure or heart rate monitors, and continuous monitoring systems capable of detecting disorders like dementia and notifying medical professionals of events such as falls.

The following are some sensors used for remote health management and their functions:

- *Sweat Sensors*: These sensors allow continuous, real-time, noninvasive detection of sweat analytes. They are directly laminated on to the human epidermis to sense chemical elements like lactate, sodium, and chloride ions in real time. Wearable devices make use of these sensors.
- Respiration Sensors: These are girth sensors that can be worn comfortably with a band. They recognize chest or abdominal expansion or contraction and produce the breathing wave pattern.
- *Blood Pressure Sensors*: These sensors detect irregular heartbeats and use "Raspberry pi" as a gateway to view the values of blood pressure online.
- *Accelerometer Sensors*: These sensors facilitate the measurement and monitoring of a person's physiological and physical activity. They can be fixed in outfits, shoes, or headbands or attached directly to the body.
- *RFID Sensors*: Radio frequency identification sensors help medical staff to pinpoint the location of any patient in the hospital to ensure their safety. These sensors are also used in automated intraoperative instrument tracking.
- *Sensor-Enabled Pills*: These refer to a pharmaceutical dosage type in which an ingestible sensor is embedded within a tablet. Some examples include imaging capsules, electrochemical sensing capsules, and gas sensing capsules.
- *EEG Sensors*: Electroencephalogram sensors can identify irregularities in the electrical activity of the brain or in human brain waves. During the procedure, electrodes made of tiny metal discs and delicate wires are applied to the scalp. The electrodes are able to detect the tiny electrical charges that brain activity produces.

- *EMG Sensors*: Electromyogram sensors detect electrical activity from a muscle using a conductive pad placed on the skin. Each muscle fiber experiences electrical impulses every time the muscle becomes active, which causes it to contract.

Advantages

- Enhances clinical decision-making using data.
- Decreases the cost of health care and increases output.
- Prevents the spread of infectious diseases and hospital-acquired infections.
- Improves patients' experience and satisfaction and also improves the clinician-patient relationship.
- It reduces readmission rates.

Disadvantages

- Accessibility and connectivity obstacles with patients.
- It is dependent on technology that not all patients can afford.
- It is totally dependent on the internet.

8.11.5 Sleep Activity Monitoring

The technique for observing a person's sleep, most often by detecting inactivity and movements, is known as sleep monitoring. In simple words, a sleep tracker is a device that keeps track of a person's sleep.

Devices such as smartphones, fitness trackers, rings, smartwatches, and other wearable devices are capable of monitoring sleep activity. Some sleep trackers are capable of recording a person's sleep stages, length of sleep, and sleep quality.

A person's motion is often measured using a gyroscope or accelerometer to establish what stage of the sleep cycle they are in. Heart rate decreases while sleeping and also varies accordingly in different stages of sleep.

These devices, equipped with a highly sensitive accelerometer sensor, determine when a person is sleeping. When the device is placed in their bed, it can record their movements throughout the night. Muscles are suppressed during deep sleep, resulting in the sleep graph becoming practically flat during this stage.

The tools utilized include the LM393 for measuring the intensity of snoring, the A&D body precision scale for measuring physical activity, the heart rate PM 235 for measuring physiological activity, the MEMS sensor HTS221 for measuring the environment of sleep, and the Fitbit Flex Bangle for measuring the level of sleep.

8.11.5.1 How Accurate Are Sleep Monitoring Systems?

The accuracy of a sleep monitor can vary based on the technology used and the biometric information gathered. For example, an EEG sensor on a sleep monitoring headset will be able to provide more data and precision than a watch that measures heart rate.

Advantages

- One significant advantage of sleep trackers and sleep sensors is that they raise awareness of the significance of sleep and when it is not being obtained.
- They can help improve sleeping habits and provide vital information to a person's doctor.
- They can also be used to rule out specific problems, such as sleep apnea, which can lead to exhaustion, hypertension, and even type 2 diabetes.

8.11.6 Real-Time Cardiac Activity Monitors

A continuous oscillatory beat is maintained in the heart via intricate electrical activity. The atrioventricular (AV) node and the sinoatrial (SA) node, which are the two primary auto-rhythmic pacemakers, produce each beat. Systole and diastole are the two primary phases of each heart cycle.

The heart's rate is a crucial factor to take into account when examining its signals. Electrodes could be positioned to track the beating of the heart. Electrodes made of silver chloride are utilized to assess ECGs in all clinical contexts. In the event of remote consultation, the Doppler effect is utilized to examine the cardiac muscles. By measuring the change in blood volume and utilizing photoplethysmography to see if there are any obstructions in the blood, it is possible to determine whether the heart is pumping blood effectively. In order to interpret light using a photodiode, a light-emitting diode (LED) is placed at remote locations during the photoplethysmography method, such as the tip of a finger or the cartilage of the ear. An implanted phonocardiographic (PCG) sensor can also measure heart rate.

Real-time cardiac monitors (RTCMs) are portable electronic devices that process ECG data quickly and conveniently online using pre-programmed algorithms. The main flexible sensors for cardiovascular vital signs monitoring based on mechanical techniques are strain sensors and pressure sensors, which can be monitored in a wearable or implanted form. A person's heart rhythm is recorded for up to 30 days while wearing a mobile cardiac monitor. Their doctor receives the results immediately after submission. This data is used by the doctor to assess symptoms and identify the arrhythmia's root cause.

8.11.7 Holter Monitor

A Holter monitor is a type of portable electrocardiogram (ECG). It continuously tracks the electrical activity of the heart for 24 hours or longer.

A typical or "resting" ECG is one of the simplest and quickest tests performed to examine the heart.

The two main uses of Holter monitors are as follows:

- To keep track of any abnormalities in heart rhythm.
- To keep an eye out for instances when the coronary arteries are not providing enough blood flow to the heart muscle.

Advantages

- Measure various aspects of heart function to assist a doctor in understanding a patient's overall heart health.
- Arrhythmias (irregular heartbeat) and the activities that have an impact on heart rate can be detected with remote cardiac monitoring.

Disadvantages

- In sports involving vigorous hand movement or flexing of muscles and tendons close to the sensor, accuracy isn't always reliable.
- There are limitations to the reliability of heart rate monitoring on dark or tattooed skin.
- A chest strap heart rate monitor is more likely to be worn improperly.

8.12 CONCLUSION

The integration of Body Sensor Networks with the Internet of Medical Things has ushered in a transformative era in health care. This synergy has brought forth a multitude of innovative applications that significantly enhance patient care, diagnostics, treatment, and overall well-being.

The seamless connectivity offered by BSNs and IoMT has revolutionized remote patient monitoring, allowing healthcare providers to remotely track patients' vital signs, physiological parameters, and health trends in real time. This has proven invaluable for individuals with chronic conditions, with post-operative care needs, and who require continuous health assessment.

Early disease detection, facilitated by the continuous data collection and analysis capabilities of BSNs in conjunction with IoMT, has opened new avenues for proactive health care. The ability to identify subtle changes and deviations from normal physiological patterns empowers medical professionals to intervene early, potentially preventing the progression of diseases and improving treatment outcomes.

Personalized medicine has taken a giant leap forward with the utilization of BSNs and IoMT. The availability of comprehensive, real-time health data enables healthcare providers to tailor treatment plans, medications, and interventions to each patient's unique needs. This patient-centric approach not only enhances the efficacy of medical care but also increases patient engagement and satisfaction.

The rise of telemedicine and telehealth owes much to the capabilities of BSNs and IoMT. Virtual consultations, diagnostics, and prescription adjustments have become more accessible and effective, bridging geographical gaps and bringing healthcare services to remote or underserved areas.

On a broader scale, BSNs and IoMT contribute to a healthier society by promoting wellness and preventive health care. Individuals can actively monitor their physical activity, sleep patterns, and overall health, leading to informed lifestyle choices and improved health outcomes.

In emergency situations, the integration of BSNs and IoMT offers life-saving potential. Swift response times, accurate patient data, and location tracking combine to enhance emergency medical services and disaster management, ultimately saving lives.

However, challenges remain, including data privacy and security concerns, interoperability issues, and the need for streamlined regulations. As the field continues to evolve, addressing these challenges will be paramount to fully realizing the potential of BSNs and IoMT in health care.

In conclusion, the application of BSNs within the framework of IoMT has reshaped health care into a proactive, patient-centered, and data-driven ecosystem. The marriage of wearable sensors, real-time data transmission, and advanced analytics holds promise for revolutionizing healthcare delivery, improving patient outcomes, and advancing our understanding of human health. As this dynamic field continues to advance, it is poised to drive innovation and shape the future of health care for generations to come.

REFERENCES

1. www.researchgate.net/figure/Architecture-for-Internet-of-Medical-Things-IoMT_fig1_334098654
2. Gravina, R., & Fortino, G. (2021, June 1). Wearable Body Sensor Networks: State-of-the-Art and Research Directions. *IEEE Sensors Journal* 21(11), 12511–12522. https://doi.org/10.1109/JSEN.2020.3044447
3. Lee, S., Ha, G., Wright, D., Ma, Y., Sen-Gupta, E., Haubrich, N., Branche, P., Li, W., Huppert, G., Johnson, M., Mutlu, B., Li, K., Sheth, N., Wright, J., Huang, Y., Mansour, M., Rogers, J., & Ghaffari, R. (2018). Highly Flexible, Wearable, and Disposable Cardiac Biosensors for Remote and Ambulatory Monitoring. *NPJ Digital Medicine*. https://doi.org/10.1038/s41746-017-0009-x. (Figure 3).
4. www.hexoskin.com/pages/health-research

Chapter 9

Home Automation and Security Systems Using AI and IoT

Ihtiram Raza Khan, Ayush Kumar Jha, Naved Ahmad, and Himanshu Rawat

9.1 INTRODUCTION

9.1.1 Background and Overview

The field of home automation and security systems has undergone remarkable advancements with the integration of artificial intelligence (AI) and Internet of Things (IoT) technologies. These systems are designed to automate and optimize various functions within a home, including lighting, temperature control, entertainment, and security, with the ultimate goal of improving convenience, energy efficiency, and safety.

Over the years, home automation has evolved from basic remote control systems to sophisticated smart homes that can be managed through mobile devices and voice commands. AI and IoT have played a pivotal role in this transformation, allowing homes to become intelligent, interconnected spaces that cater to the personalized needs of residents.

By leveraging AI algorithms, smart home devices can learn user preferences and adapt to their behavior, providing a seamless and tailored experience. Additionally, the integration of IoT enables real-time data exchange and analysis, allowing homeowners to monitor and control their homes remotely.

Overall, the integration of AI and IoT in home automation and security systems has revolutionized the way we interact with our living spaces, making our homes more efficient, secure, and user-friendly.

9.1.2 Objective

The objective of implementing home automation and security systems using AI and IoT is to create smart and interconnected living spaces that enhance convenience, energy efficiency, and security for homeowners. By leveraging AI algorithms and IoT technologies, the aim is to automate various aspects of home management, such as lighting, temperature control, entertainment, and access control, to provide a seamless and personalized user experience.

DOI: 10.1201/9781003406723-9

The integration of AI and IoT in home automation systems seeks to optimize energy consumption, reduce waste, and lower utility costs by enabling intelligent control of devices based on user behavior and environmental factors. Additionally, these systems aim to enhance security through AI-powered surveillance cameras, facial recognition, and smart access control solutions, ensuring the safety of residents and their property.

Overall, the objective is to transform traditional homes into smart homes that adapt to user preferences, offer remote monitoring and control capabilities, and provide an elevated level of security, ultimately improving the quality of life for homeowners.

9.1.3 Scope and Limitations

9.1.3.1 Scope of Home Automation and Security Systems using AI and IoT

The integration of AI and IoT has transformed home automation and security systems, offering a wide array of possibilities. These systems comprise various smart devices, including lighting, thermostats, locks, cameras, entertainment systems, and AI-driven virtual assistants. AI algorithms enable devices to learn user preferences, providing personalized experiences. IoT connectivity facilitates seamless data exchange among devices, creating an intelligent and interconnected home environment. Automation optimizes lighting, temperature, and entertainment, enhancing energy efficiency and convenience. AI-powered surveillance systems bolster home security by detecting threats, while remote monitoring ensures homeowners' peace of mind. AI-driven virtual assistants offer intuitive interfaces and voice commands, creating a user-friendly living environment that responds to residents' preferences.

9.1.3.2 Limitations of Home Automation and Security Systems Using AI and IoT

The integration of AI and IoT technologies in home automation and security systems brings significant benefits, but it also comes with several limitations and challenges that stakeholders must address. The initial cost of implementing these advanced technologies can be substantial, especially for comprehensive solutions. Data privacy and security concerns arise as smart devices collect sensitive information, necessitating robust measures to protect personal data. Interoperability challenges may arise due to different devices' incompatibility, limiting seamless integration. Ensuring the reliability of AI algorithms and IoT connectivity is crucial for consistent performance, requiring constant monitoring and updates.

Setting up and maintaining a smart home system may be complex, posing challenges for non-tech-savvy users. User acceptance and adaptation to new technologies also need consideration, as some individuals may struggle to embrace changes and alter their habits. Rapid technological advancements may lead to compatibility issues with older devices. The reliance on internet connectivity introduces network vulnerabilities, requiring robust security measures.

The use of AI-driven surveillance and data analytics raises ethical and social concerns regarding privacy and surveillance. Addressing these ethical implications is essential for the responsible deployment of AI and IoT in home automation and security systems.

Understanding the scope and limitations of AI and IoT in home automation and security is crucial for stakeholders to make informed decisions and maximize benefits while mitigating challenges effectively. By embracing advancements responsibly, stakeholders can shape a future where intelligent and secure homes enhance lives while respecting privacy and ethical considerations.

9.2 METHODOLOGY

9.2.1 Data Collection

Home detection and security systems (HDSS) are becoming increasingly popular as they offer a way to improve the safety and security of homes. These systems typically collect data from a variety of sensors, such as motion detectors, door and window sensors, and cameras. This data is then used to detect unauthorized access, fire, and other hazards.

The types of data collected by HDSS include the following:

- Sensor data, device data, and user data. Sensor data is collected from sensors such as motion detectors, door and window sensors, and temperature sensors. This data can be used to track movement in and around the home, as well as to detect changes in temperature or humidity.
- Device data is collected from devices such as lights, locks, thermostats, and security cameras. This data can be used to track the status of these devices, as well as to record events such as when a door is opened or closed.
- User data is collected from users, such as their login information, usage patterns, and preferences. This data can be used to personalize the HDSS experience, as well as to track the effectiveness of the system.

The purpose of data collection in HDSS is to improve the safety and security of homes, as well as to personalize the user experience. By collecting data, HDSS can identify patterns and trends, which can be used to optimize the system's performance. For example, data on how often a user opens and closes a door can be used to optimize the system's energy efficiency.

Additionally, data collection can be used to personalize the user experience by learning the user's preferences for lighting and temperature and adjusting the settings accordingly.

The security of data collected by HDSS is of paramount importance. This data can be used to identify and track users, and it could also be used to gain access to a home's security system. Therefore, it is important that these systems use strong encryption and other security measures to protect user data.

Overall, data collection is a valuable tool for improving the performance and security of HDSS. However, it is important to be aware of the potential privacy and security risks associated with data collection.

9.2.1.1 Primary Data

Primary data is data that is collected for the first time, specifically for the purpose of the study. It can be collected through surveys, interviews, observational studies, or experiments. Primary data can be a valuable source of information for understanding how home automation and security systems are used and for identifying areas for improvement.

The benefits of collecting primary data include more accurate data, more relevant data, and new insights. However, it is important to be aware of the potential drawbacks before collecting primary data, such as the time and expense involved.

The following are some examples of primary data that could be collected in home automation and security systems:

- User preferences for lighting, temperature, and security settings
- Usage patterns
- Problems and challenges
- Security concerns

This data can be used to improve the performance and security of home automation and security systems and to personalize the user experience.

9.2.1.2 Secondary Data

- Secondary data can be a valuable source of information for researchers on home automation and security systems.
- There are many different sources of secondary data, including government data, industry data, academic research, and news and media.
- Secondary data can provide insights into trends, demographics, market size, growth rates, and other factors that can affect the development and use of home automation and security systems.
- However, it is important to be aware of the limitations of secondary data, such as its accuracy, timeliness, and potential biases.

9.2.2 Research Design

Research design is an important consideration for any study of home automation and security systems using AI and IoT. The specific data collection and analysis methods will depend on the research questions being asked, but some general considerations include the following:

- Data collection: Researchers may collect data from sensors, user behavior, or other sources. It is important to consider the privacy implications of data collection, as well as the ethical implications of using AI and IoT in research.
- Data analysis: Researchers may use a variety of methods to analyze data, such as statistical analysis, machine learning, and natural language processing. The specific method will depend on the research questions being asked and the type of data collected.
- Research ethics: Researchers must ensure that the privacy of users is protected and that the research does not cause any harm to the users. This is especially important when using AI and IoT, as these technologies can collect and analyze a vast amount of data about users [1].

9.2.3 Home Automation Architecture

Home automation architecture is the design of the underlying infrastructure that allows devices and systems to communicate with each other and with users. AI and IoT are two key technologies that are driving the development of home automation architecture.

9.2.3.1 Sensing and Data Collection

Sensing and data collection are essential components of home automation and security systems. Sensors are used to collect data about the environment, such as temperature, humidity, light levels, and motion. This data can then be used to automate tasks, such as turning on lights when it gets dark and locking the doors when someone leaves the house.

There are a variety of sensors that can be used in home automation and security systems. Some of the most common sensors include motion sensors, temperature sensors, humidity sensors, gas sensors, and smoke sensors. Data collection is the process of gathering data from sensors. This data can be collected in real time or stored for later analysis. The type of data collected will depend on the specific needs of the home automation or security system.

In short, sensing and data collection are essential for home automation and security systems. By collecting data from sensors, these systems can be made more efficient, secure, and user-friendly [2].

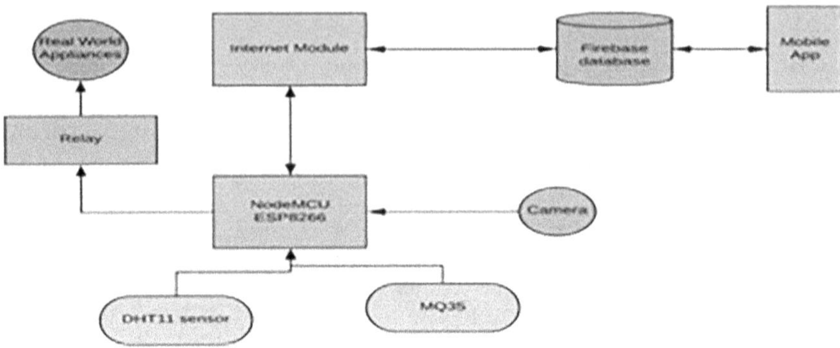

Figure 9.1 Process of home automation using sensors.

9.2.3.2 Communication Protocols

Communication protocols are the rules that govern how devices communicate with each other. There are a number of different communication protocols that can be used in home automation and security systems. Some of the most common protocols include the following:

- Zigbee: Zigbee is a low-power, wireless mesh networking protocol that is well-suited for home automation and security applications. Zigbee is used by a wide range of devices, including light bulbs, thermostats, and security cameras.

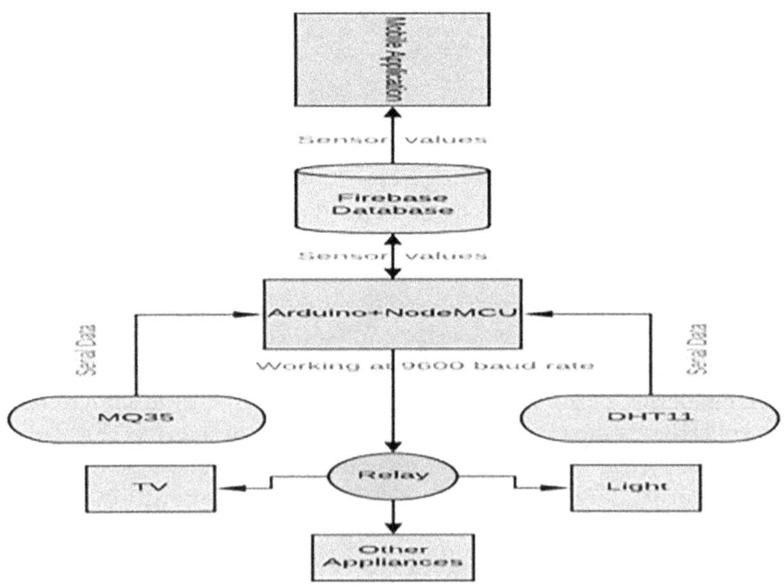

Figure 9.2 IoT integration in home automation.

- Z-Wave: Z-Wave is another low-power, wireless mesh networking protocol that is popular for home automation and security applications. Z-Wave is similar to Zigbee, but it uses a different radio frequency.
- Wi-Fi: Wi-Fi is a high-speed, wireless networking protocol that is commonly used for home automation and security applications. Wi-Fi is a good choice for devices that require high bandwidth, such as video cameras.
- Ethernet: Ethernet is a wired networking protocol that is often used for home automation and security applications. Ethernet is a good choice for devices that require high reliability and security, such as security cameras.

The choice of communication protocol will depend on the specific needs of the home automation or security system [3].

9.2.4 AI and IoT Integration in Home Automation and Security

Home automation has evolved significantly over the years, from basic timers and remote controls to complex systems that integrate multiple devices and technologies. AI and IoT have made home automation more intelligent and efficient. These technologies are changing the way we live in our homes, providing higher levels of comfort, security, and energy efficiency.

IoT sensors and devices collect data that is analyzed by AI algorithms, allowing the system to learn from users' preferences. For example, an AI algorithm can determine when a person typically wakes up and adjust the temperature and lighting accordingly. This results in a more personalized and effective home automation system.

IoT devices such as smart locks, security cameras, and thermostats can be integrated into a home automation system to improve security and monitoring. Users can remotely monitor their home and receive alerts if anything unusual happens. This not only provides peace of mind, but also allows users to take action quickly if necessary.

9.2.4.1 Applications of AI in Home Automation and Security System

The application of AI in home automation systems can be divided into four categories: comfortable systems, remote controlling systems, resource optimizing systems, and secure systems. In comfortable systems, AI is used to create a knowledge-based database that can be learned by the system to improve comfort. In remote controlling systems, AI can be used to authorize commands and make decisions about which devices to control. In resource optimizing systems, AI can be used to analyze data and make recommendations for improving efficiency. In secure systems, AI can be used to process video, image, and audio data to detect security threats.

Overall, AI has a wide range of applications in home automation systems. As AI technology continues to develop, it is likely that AI will become even more important in the future of home automation.

The following are some specific examples of how AI is used in home automation systems:

- In comfortable systems, AI can be used to learn the user's preferences for temperature, lighting, and other settings. This allows the system to automatically adjust these settings to create a more comfortable environment.
- In remote controlling systems, AI can be used to authorize commands from remote devices. This can be used to control home appliances, security systems, and other devices from anywhere in the world.
- In resource optimizing systems, AI can be used to analyze data and make recommendations for improving efficiency. For example, AI can be used to identify patterns in energy usage and recommend ways to reduce energy consumption.
- In secure systems, AI can be used to process video, image, and audio data to detect security threats. This can be used to protect homes from intruders, burglars, and other threats [4].

9.3 BENEFITS AND CHALLENGES OF AI AND IoT IN HOME AUTOMATION AND SECURITY

9.3.1 Benefits

9.3.1.1 Enhanced Convenience and Comfort

AI and IoT technologies have revolutionized home automation, making daily tasks more convenient and effortless for residents. With these technologies, routine actions such as adjusting lighting, regulating temperature, and controlling entertainment systems can be automated and managed seamlessly. Homeowners can now enjoy the luxury of controlling their home devices using voice commands or smartphone apps, eliminating the need for manual interventions.

For example, one can simply instruct a virtual assistant to dim the lights or set the thermostat to a preferred temperature without having to physically interact with switches or thermostats. This level of automation creates a comfortable and intuitive living environment where the home adapts to the residents' preferences and needs.

In essence, AI and IoT technologies in home automation provide unparalleled convenience, transforming traditional homes into intelligent spaces that cater to residents' desires, ultimately enhancing the overall living experience.

9.3.1.2 Energy Efficiency and Savings

Smart home systems integrated with AI algorithms have the capability to optimize energy consumption by analyzing user behavior and environmental conditions. Through continuous data collection and analysis, these systems learn the patterns and preferences of the residents. Based on this knowledge, the smart devices autonomously adjust various settings, such as thermostat schedules and lighting levels, to ensure energy efficiency and cost savings.

For example, if the system detects that a particular room is unoccupied during certain hours, it can automatically adjust the thermostat to reduce heating or cooling, thereby saving energy. Similarly, the smart lighting system can dim or turn off lights in rooms that are not in use, contributing to further energy conservation.

Over time, as the AI algorithms gather more data and refine their understanding of user habits, the energy optimization becomes more accurate and effective. As a result, homeowners experience significant energy savings, leading to reduced utility costs and a more environmentally sustainable living environment. The combination of AI-driven optimization and IoT connectivity allows smart home systems to adapt and respond intelligently to the changing needs and behaviors of residents, ensuring a comfortable and energy-efficient home.

9.3.1.3 Improved Security and Safety

AI-powered surveillance cameras and smart access control systems play a crucial role in enhancing home security by leveraging advanced technologies such as AI. These surveillance cameras are equipped with AI algorithms that enable real-time monitoring and analysis of the surroundings. The cameras can detect and recognize faces and objects, allowing them to identify potential threats or intruders.

With facial recognition technology, the surveillance cameras can distinguish between known individuals (such as family members or authorized visitors) and unknown individuals, triggering alerts when unidentified people are detected. This capability enhances the safety of residents by providing immediate notifications in case of any suspicious activities.

Moreover, the cameras' object detection feature enables them to identify and track objects, such as packages, bags, or vehicles, within the monitored area. This helps to prevent theft or unauthorized entry, as any suspicious object detected can prompt immediate action or alert the homeowners.

9.3.1.4 Personalization and Customization

AI and IoT technologies empower smart homes to learn and understand residents' preferences through data analysis and machine learning algorithms. These systems can collect and analyze data about users' behaviors, habits, and interactions with smart devices. By identifying patterns and trends, AI algorithms can create personalized profiles for each household member.

Based on these profiles, smart homes can adapt their functionalities to cater to individual preferences. For example, the lighting system can automatically adjust brightness and color temperature according to a resident's preferred settings. The thermostat can learn the preferred temperature settings for different times of the day and adjust accordingly to maintain comfort.

This level of personalization allows each household member to enjoy a unique and tailored experience within the smart home environment. As residents interact with smart devices over time, the AI algorithms continually refine and update their understanding, further enhancing the personalized features and optimizing the overall user experience.

9.3.1.5 Remote Monitoring and Control

With the integration of AI and IoT technologies in home automation and security systems, homeowners can enjoy the convenience of remote access capabilities. This means they can easily monitor and control their smart home devices from anywhere using their smartphones or other connected devices.

For example, if a homeowner is at work or on vacation, they can access their smart home system through a dedicated mobile app. From the app, they can adjust the thermostat to ensure the house is comfortable before they return, turn on or off lights to create the illusion of someone being home, or even check the security cameras to see if everything is okay.

This feature offers flexibility and peace of mind to homeowners, knowing that they have control over their home even when they are not physically present. It enhances the overall home management experience, providing a sense of security and convenience. Remote access allows homeowners to stay connected to their smart home ecosystem and make necessary adjustments in real time, ensuring their comfort and safety regardless of their location.

9.3.2 Challenges

9.3.2.1 Data Privacy and Security Risks

The widespread use of interconnected devices in home automation and security systems raises valid concerns about data privacy and potential security risks. Smart home devices collect and exchange sensitive information, such as personal preferences, routines, and usage patterns, which can be susceptible to unauthorized access or misuse.

To safeguard homeowners' privacy, it is crucial for manufacturers and service providers to implement robust security measures. This includes strong encryption protocols to protect data during transmission and storage, secure authentication methods to ensure authorized access, and regular software updates to address potential vulnerabilities.

Additionally, homeowners should be vigilant in configuring their smart devices securely, such as using strong passwords, enabling two-factor authentication, and regularly reviewing permissions granted to third-party applications or services.

Addressing data privacy and security concerns is vital to ensure that smart homes remain a safe and secure environment for residents, fostering trust in the adoption of AI and IoT technologies.

9.3.2.2 Interoperability and Standardization

Smart home systems integrated with AI algorithms can analyze user behavior and environmental conditions to optimize energy usage. By continuously learning from user interactions, these systems can intelligently adjust settings, such as thermostat schedules and lighting levels, to ensure energy efficiency without compromising comfort. For example, the smart thermostat can learn the residents' preferred temperature patterns and adjust heating or cooling accordingly, minimizing energy waste.

Moreover, the AI-powered smart lighting system can automatically adjust brightness levels based on natural light availability and occupancy, further reducing energy consumption. These optimizations lead to significant energy savings over time, resulting in reduced utility costs for homeowners.

By leveraging AI to make data-driven decisions, smart home systems create a more sustainable and environmentally friendly living space. Homeowners benefit from a more efficient and cost-effective home environment while contributing to global efforts to conserve energy and reduce carbon footprints.

9.3.2.3 Reliability and Dependability

For smart home systems to operate reliably, the performance of AI algorithms and IoT connectivity is crucial. AI algorithms are responsible for processing and analyzing data collected from various smart devices, enabling intelligent decision-making and automation. Likewise, IoT connectivity facilitates seamless communication and data exchange among these devices, forming an interconnected network within the home.

Regular updates and monitoring are essential to ensure the continued dependability of these systems. AI algorithms may need updates to improve their functionality, enhance security, or adapt to changing user preferences. Similarly, IoT connectivity requires monitoring to detect any potential issues or disruptions that could impact the system's performance.

By maintaining consistent and efficient operation through regular updates and monitoring, smart home systems can provide homeowners with the convenience, security, and energy efficiency they desire. Reliable AI algorithms and IoT connectivity contribute to a seamless and user-friendly smart home experience, promoting widespread adoption and satisfaction among users.

9.4 FUTURE ENHANCEMENTS

As technology continues to advance, the integration of AI and IoT into home automation and security systems has become increasingly prevalent. AI and IoT technologies offer numerous benefits, including increased convenience, enhanced security, and improved energy efficiency. In this section, we will explore three key areas for future enhancements in home automation and security systems: advanced AI algorithms, predictive maintenance, contextual awareness, and emotional intelligence in AI assistants.

9.4.1 Advanced AI Algorithms

9.4.1.1 Predictive Maintenance

AI algorithms can be utilized to predict and prevent potential issues in home automation and security systems. By analyzing historical data and monitoring system performance, predictive maintenance algorithms can identify patterns and anomalies, allowing for proactive maintenance. This approach helps to minimize downtime, reduce repair costs, and enhance the overall reliability of the systems.

9.4.1.2 Contextual Awareness

Contextual awareness refers to the ability of AI systems to understand and interpret the context in which they operate. In the context of home automation and security systems, contextual awareness enables the AI algorithms to adapt and respond intelligently to different situations. For example, an AI-enabled security system can differentiate between a family member entering the house and an intruder, based on factors such as facial recognition, behavior patterns, and time of entry. This level of contextual awareness enhances the accuracy and effectiveness of the security system.

9.4.1.3 Emotional Intelligence in AI Assistants

Emotional intelligence in AI assistants refers to their ability to recognize and respond appropriately to human emotions. By incorporating emotional intelligence into AI assistants, home automation systems can provide a more personalized and empathetic user experience. For instance, an AI assistant can detect if a user is feeling stressed or anxious and adjust the lighting or temperature or play soothing music to create a calming environment. This level of emotional intelligence enhances the overall user satisfaction and comfort within the home.

9.4.2 Seamless Device Interoperability

Seamless device interoperability refers to the ability of different smart devices within a home automation system to communicate and work together seamlessly. It ensures that various devices from different manufacturers can interact with each other, exchange data, and collaborate in a harmonious manner. This interoperability is essential to create a unified and integrated smart home ecosystem, where devices can collaborate to provide enhanced functionalities and a smoother user experience.

9.4.2.1 Common Communication Standards

One of the key challenges in the early stages of home automation was the lack of standardized communication protocols among different smart devices. Each manufacturer often implemented proprietary communication methods, making it challenging for devices to interact with those from other brands. This lack of standardization led to compatibility issues and limited choices for consumers, and it made the setup and integration of smart home devices more complex.

To address this issue, the industry has been working towards establishing common communication standards. Standardized protocols, such as Wi-Fi, Bluetooth, Zigbee, and Z-Wave, enable smart devices to communicate using a universal language, ensuring seamless interoperability. With common communication standards in place, users can easily add new devices to their existing smart home network, as these devices can "speak the same language" and understand each other's commands and data.

9.4.2.2 Integration with Emerging Technologies (e.g., 5G)

The integration of smart home systems with emerging technologies like 5G is another crucial aspect of seamless device interoperability. 5G is the fifth generation of cellular technology, offering significantly higher data speeds, lower latency, and increased capacity compared to its predecessors. When integrated with smart home devices, 5G enables faster and more reliable communication between devices and the cloud, enhancing overall system performance.

With 5G's low latency and high bandwidth capabilities, smart home devices can transmit and receive data in real time, facilitating quick responses to user commands and enabling near-instantaneous communication between devices. This leads to a more responsive and efficient smart home experience. Additionally, 5G's increased capacity can support a higher number of connected devices, allowing for the seamless integration of multiple smart devices within a home.

Moreover, 5G connectivity can extend beyond the home environment, allowing homeowners to remotely monitor and control their smart devices from anywhere with minimal delay. This enhanced connectivity opens up new possibilities for smart home applications, enabling real-time monitoring of security cameras, smart thermostats, and other devices on the go.

9.4.3 Edge Computing for Enhanced Privacy and Speed

Edge computing is an innovative paradigm that involves performing data processing and storage at the edge of the network, closer to the source of data generation. In the context of home automation and security systems using AI and IoT, edge computing brings significant advantages, particularly in terms of enhanced privacy and faster response times.

9.4.3.1 Reducing Reliance on Cloud Services

Traditionally, many smart home devices relied heavily on cloud services for data processing and storage. This meant that data generated by these devices, such as video footage from security cameras or sensor readings from various smart sensors, would be sent to a remote cloud server for analysis and storage. While cloud services offer scalability and accessibility, they also raise concerns about data privacy and security. Transmitting sensitive data to the cloud increases the risk of potential data breaches and unauthorized access.

With edge computing, data processing occurs locally on the smart devices themselves or in nearby edge computing nodes. This localized data processing reduces the need for constant data transmission to the cloud, minimizing exposure to potential security vulnerabilities. Homeowners can have greater confidence that their sensitive data remains within their local network, leading to enhanced data privacy and security.

9.4.3.2 Faster Response Times

Another significant advantage of edge computing is its ability to provide faster response times for smart home devices. When data is processed at the edge of the network, there is a substantial reduction in latency compared to sending the data to a remote cloud server for processing. This is particularly crucial for real-time interactions and critical applications in home automation and security.

For instance, in a smart security system, edge computing enables rapid analysis of video footage and immediate detection of potential threats. Instead of waiting for the data to travel to the cloud server and back, the smart cameras can analyze the data locally and provide instant alerts to homeowners about any suspicious activities. Similarly, in smart home automation, edge computing allows for quicker response times when adjusting devices like smart thermostats or lighting systems.

The reduced latency provided by edge computing enhances the overall user experience in smart homes. Homeowners can enjoy real-time control and interactions with their smart devices, making the automation and security systems more efficient and responsive [5].

9.4.4 AI and Blockchain for Enhanced Security

AI and blockchain are powerful technologies that can enhance security in home automation and security systems using AI and IoT. By combining AI's advanced pattern recognition and behavioral analysis with blockchain's decentralized and tamper-proof nature, decentralized authentication and access control can be achieved, reducing the risk of unauthorized access. Additionally, storing surveillance footage on a blockchain ensures its immutability, providing reliable and tamper-proof evidence for smart home security, making homes safer and more secure.

9.4.4.1 Decentralized Authentication and Access Control

The combination of AI and blockchain technology offers promising solutions for enhanced security in home automation and security systems. Traditional centralized authentication methods, such as passwords and biometrics stored on a central server, can be vulnerable to hacking and unauthorized access. Decentralized authentication leverages the blockchain's distributed ledger technology and AI's advanced pattern recognition capabilities to create a more secure and tamper-resistant authentication process.

In a decentralized authentication system, user credentials and access permissions are securely stored on a blockchain network. Each user has a unique digital identity, represented by cryptographic keys. AI algorithms can analyze user behavior patterns and device interactions to continuously assess and validate user identities. This enables a dynamic and adaptive authentication process, where access is granted or revoked based on real-time behavioral analysis.

Decentralized authentication reduces the risk of single points of failure and unauthorized access attempts. The distributed nature of the blockchain network ensures that compromising one node does not compromise the entire system. As a result, home automation and security systems become more resilient against security breaches, safeguarding sensitive data and preventing unauthorized individuals from gaining access to smart home devices.

9.4.4.2 Tamper-Proof Surveillance Footage

Blockchain's tamper-proof nature also finds application in securing surveillance footage in smart home security systems. Traditional centralized storage systems may be susceptible to tampering or deletion of recorded footage, compromising the reliability and integrity of evidence in case of security incidents or legal investigations.

By storing surveillance footage on a blockchain network, the data becomes immutable and tamper-proof. Each recorded video segment is encrypted and timestamped, forming a chain of blocks linked through cryptographic hashes. Once data is recorded and added to the blockchain, it cannot be altered or deleted without consensus from the entire network.

This tamper-proof nature ensures the reliability and authenticity of surveillance footage, making it admissible as evidence in legal proceedings. In case of security incidents, homeowners and law enforcement can rely on the integrity of the recorded footage to investigate and resolve incidents accurately.

Moreover, the combination of AI and blockchain in surveillance systems can enhance video analytics capabilities. AI-powered video analysis algorithms can process the surveillance footage, enabling advanced features like facial recognition, object detection, and anomaly detection. This empowers smart security systems to detect and respond to potential threats more effectively, further enhancing home security.

9.5 CONCLUSION

Home automation and security systems have come a long way in recent years, thanks to the advances in artificial intelligence and the Internet of Things. These technologies are now being used to create homes that are more comfortable, secure, and energy-efficient.

In this chapter, we have explored the many ways that AI and IoT can be used in home automation and security systems. We have discussed the different subtopics in this chapter, including the following:

- Sensors and actuators: These devices are used to collect data about the environment and to control devices in the home.
- Communication protocols: These protocols allow devices to communicate with each other and with the cloud.
- AI algorithms: These algorithms are used to analyze data and to make decisions about how to control devices in the home.
- Security and privacy: These are important considerations when designing home automation and security systems.

We have also looked at some of the challenges that need to be addressed in order to make home automation and security systems more widespread. These challenges include the following:

- Cost: Home automation and security systems can be expensive to purchase and install.
- Complexity: Home automation and security systems can be complex to design and install.
- Security: Home automation and security systems need to be secure from cyberattacks.

Despite these challenges, the future of home automation and security systems is bright. As AI and IoT technology continues to develop, these systems will become more affordable, easier to use, and more secure. As a result, we can expect to see a significant increase in the use of home automation and security systems in the years to come.

We hope that this chapter has given you a better understanding of the potential of AI and IoT in home automation and security systems. We believe that these technologies have the power to make our homes more comfortable, secure, and energy efficient. We encourage you to learn more about these technologies and to consider how they could be used to improve your own home.

REFERENCES

1. "A Review on IoT Based Smart Security and Home Automation" by S. M. Nasir and M. H. M. Rashid (2022).
2. "A Sensor Based Home Automation and Security System" by S. S. B. Jadhav and M. S. Patil (2016).
3. "Smart Home Automation System Using IoT, AI and Communication Protocols" by Kiran Kumar P. Johare, Vasant G. Wagh and Arvind D. Shaligram (2022).
4. "Application of AI in Home Automation" by Sandeep Kumar and Mohammed Abdul Qadeer (2012).
5. "Edge Computing Framework for Enabling Situation Awareness in IoT Based Smart City" by S. Alamgir Hossain, Md. Anisur Rahman and M. Anwar Hossain (2018).

For Product Safety Concerns and Information please contact our EU representative GPSR@taylorandfrancis.com
Taylor & Francis Verlag GmbH, Kaufingerstraße 24, 80331 München, Germany

www.ingramcontent.com/pod-product-compliance
Lightning Source LLC
LaVergne TN
LVHW010604110826
845149LV00003B/767

* 9 7 8 1 0 3 2 5 2 4 4 8 1 *